T0003851

my cool classic car.

TRANS AM

my cool classic car.

an inspirational guide to classic cars

chris haddon

PAVILION

AAFES
LITRES
UNLEADED
15
UNLEADED

contents

introduction ... 6

beloved ... 10
vw beetle ... 12
fiat 500 ... 16
morris minor ... 20
austin seven 850 ... 24
renault 4 ... 28
citroën 2cv ... 32
austin 7 chummy ... 36
morgan 4/4 ... 38
triumph herald ... 44

retro ... 46
vw golf ... 48
rover sd1 3500 vitesse ... 52
austin allegro vanden plas ... 54
hg holden ute ... 56
amc pacer ... 58
rover p6 ... 62
toyota celica gt2000 ... 64
ford capri mk1 ... 68
fiat x1/9 ... 72
panhard 24ct ... 74

glory days ... 78
bentley r type continental ... 80
corvette v8 speedster 'duntov' ... 82
plymouth belvedere ... 84
jaguar e-type ... 88
facel vega ... 92
ford mustang ... 96
pontiac firebird trans am ... 98

classic eccentrics ... 102
goggomobil ts300 coupé ... 104
tatra t97 ... 108
willys jeep ... 112
porsche 356a ... 116
delorean ... 118
morris 1800 ... 122
ford model t ... 126
citroën ds pallas ... 130

sports ... 134
mercedes 300sl roadster ... 136
mga mk2 coupé ... 142
porsche 912 ... 144
volvo p1800e ... 148
triumph stag ... 152
morgan 3 wheeler ... 154

sourcebook ... 156
credits ... 158
acknowledgements ... 159

introduction

As a child I remember my father returning home with a steering wheel and instrument panel from a late 1960s Rover P5. After a few hours in his workshop, triumphantly he emerged with the ultimate in-car toy. This was built for me with love, but I suspect an ulterior motive was an attempt at curtailing my 'are we there yet?' habit when only a few miles into a journey. I was the envy of my school chums, but quite what on earth my parents were thinking is another matter as it was nothing short of lethal. This was the 1970s, and very much pre-Health and Safety. No Fisher Price multicoloured soft-edge precautions with this toy! Happily, I remained unscathed and as the miles went by sitting in the back of my father's Triumph Dolomite, you would find me – much to my parents' delight – endlessly flicking the switches, turning the wheel and copying my father's every move without even a murmur about our journey's progress.

Time goes by and memories fade but technology moves on regardless and it was only recently that, by chance, I saw a Rover P5 and the memories flooded back. So strong were they that I could vividly remember the visual and tactile nature of my very own childhood instrument panel and steering wheel, this from the days when dashboards and interior spaces were art-forms in their own right. It might seem trivial to lust over something as apparently simple as a switch, but they used to activate with a satisfying 'clunk and a click' leaving you in little doubt the switch was most certainly on. Nowadays the emphasis appears to focus on emotionless touch screens that give little or no feedback to the driver to confirm that the action has been accomplished. Neither did cars have seats that folded into 40 different positions, with additional seating and storage appearing from the most unexpected places. We simply just coped, packed less and travelled with fewer passengers. Now the familiar key threatens to be removed and replaced with the option of keyless entry and push-button ignition – when did it become so difficult to use a key? These things add up to a feeling that 'progress' at times can feel like a step in the wrong direction.

Design individuality used to be a selling point, with manufacturers daring to be different. Sadly, more often nowadays this form of distinctive design has made way for familiarity by prioritising build costs, environmental and safety considerations, or just what a computer or focus group perceives as a style that should be followed. A style that now, in many cases, lacks the level of cultural distinction that renowned automotive designers and styling houses were once widely known for. Today, new cars come and go, some hardly leaving an impression – the blurred, cross-owned design DNA of many of today's marques deem them indistinguishable from each other. I'm not the only one to note these observations and often, when discussing or admiring classic cars, similar conversations will develop. Talk will turn to quirky design aspects big or small, good or bad handling traits and humorous, almost slightly haphazard, approaches to solutions. But in all cases it embraces the manufacturers' individuality and trends of the era.

I am not saying we should markedly revert to the times classic car owners cherish – the march of technology just won't allow it. However, it could be said that nowadays naive, fresh-faced drivers have missed out on the real driving experience; indeed, who knows how much of a hands-off experience driving will become in the years ahead?

So in these days of hi-tech motoring, the increasing trend for the young, old, couples, singles and families to go down the route of classic car ownership is something to be welcomed. However, such a route should not be taken lightly. A classic car can be a costly investment, maybe not in the purchase price but certainly with the ongoing maintenance. It's often down to you developing an immense connection with the car, learning everything – good and bad – that needs to be known.

But despite all the costs and hazards of classic car ownership, I want to focus on the immense positives. The desire to own a classic car is fuelled by passion, nostalgia and perhaps eccentricity, rather than necessity. With ownership comes responsibility, as it falls upon you to keep the car roadworthy – you are to some extent the temporary custodian of a piece of history, something that will be passed down through the family or on to someone new for them to cherish. Your ownership stories and workmanship will often be mentioned, judged and scrutinised by others long after you have relinquished the vehicle.

Choosing which cars to include was a difficult task. For myself, an appreciation of cars is found not in their speed and horsepower but in their flowing lines, intricate details and the sound and emotional experience. It would have been all too easy to fill the book with permanent garage-dwelling concourse cars. Instead I took the route of including some often overlooked cars, opting for those that not only embraced the owner's expression and personality, but had an interesting history to boot. Some are immaculate, while some even come with rust – above all, regardless of age and value they are regularly used, clocking up significant mileage. For the purists among you, you will notice a few cars that do break the term 'classic' and encroach into 'vintage' territory and some choices that are present-day future classics. I apologise for this – however I deemed these too important to leave out.

The career of the book's photographer, Lyndon McNeil, began as a small child while watching his father, a chief observer in a team of motor-racing marshals. Lyndon was given a camera and the rest is history. The respect that Lyndon's photographic skill commands has led him on a career path travelling the world on the F1 circuit and he is always in pursuit of cars new and old to photograph.

On our journey we have met some amazing people who have been only too happy to share their motoring experiences and explain the reasoning behind their choice of car. These owners have made this book an absolute pleasure to work on. You could say this is a sideways look at classic cars and you may well be correct. Nevertheless, I hope you enjoy the examples featured; maybe it will inspire some of you to embark on the search, like I have, for your own 'cool classic car'. And what better way to enjoy those rare moments of 'me' time than behind the wheel of a car from a different age.

beloved

Throughout life, regardless of age or gender, there are few possessions that are held in the same high regard that cars can achieve. For some their car will never have a selling price, such is the connection between owner and vehicle. It is a bond so strong that they will do all they can to avoid it becoming someone else's 'beloved' car.

'Beloved' status may be assigned to a car for any number of reasons, whether it's a classic car or not. It may be your first car affording you a first taste of freedom to explore without being reliant on others, it may hold recollections of a special person, or it may be a public expression of your individuality.

As this chapter unfolds you will read about treasured cars packed full of memories. Four-wheeled heirlooms handed down to a son or daughter so they too can add further memorable journeys to the car's already rich history before in due course it will be passed on to the next generation. A fastidious car owner, unwilling to leave anything to chance, hiding his car away where, owing to unforeseen circumstances, it remained largely forgotten for decades. A car seen and loved as a child which now, after a somewhat shaky start, is again much cherished. Obsessions developed from a chance observation. And an artist finally fulfilling his long-held dream of owning a Morgan.

No two people share exactly the same reasons that symbolise the 'beloved' status of their pride and joy, but the following cars are all worthy of the esteem in which they are held by their owners.

vw beetle

In the era of change from shillings and gallons to new pence and litres, this 1959 VW Beetle 1200 was tucked away, then locked up in the owner's garage with the key hidden for added reassurance. The car was then gently warmed by a boiler in an adjacent heated room. 'The Beetle was deliberately put into hibernation when the owner had to go and work abroad. It seemed the logical solution to keep his pride and joy dry (and, truth be told, to stop his children from taking it out for a sneaky drive),' explains present owner Lee.

Sadly, for various reasons the owner never got around to removing it from the garage, so the car remained warm and dry in its cocoon for over 35 years. When finally unearthed from under a pile of cardboard boxes, the family discovered a wonderfully preserved Beetle along with a time-capsule treasure trove. The glove box yielded up personal belongings, original receipts, a copy of *The Highway Code*, manuals and a detailed driving log. The car also contained numerous parts wrapped up in the newspapers of the time and others in their original 1970s German packaging.

Given the car's history, Lee has opted to retain its 'as-found condition' as he fears any repair or restoration works would tamper with its integrity. A much loved car has been given a new lease of life and, although it's still garaged, it doesn't stay in there for long because Lee finds that with three young daughters this appealing car is often in demand.

EV
VW

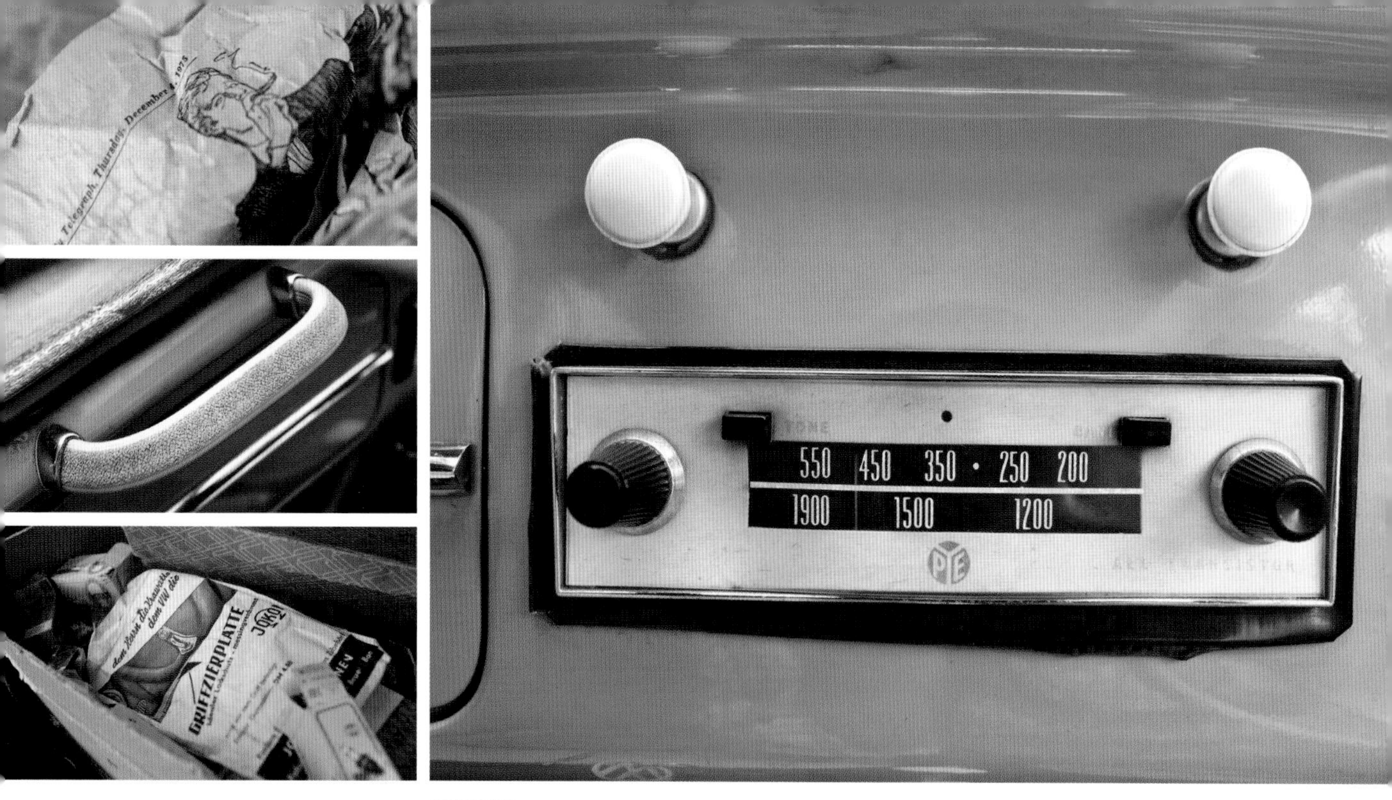

car notes

In 1938 the streamlined Volkswagen Beetle or, as it was affectionately later known, 'the bug', was designed by forward-thinking, Austrian-born car designer Ferdinand Porsche. The original brief was to develop a 'people's car' able to accommodate two adults and three children at 100km/h (62mph) on the ever-growing network of autobahns. It wasn't until after WWII that it was manufactured in significant numbers in West Germany.

During the 1950s the air-cooled, rear-engine, rear-wheel-drive Beetle was considered more refined, more economical and more powerful than most of its European counterparts. The Beetle was progressively updated over the years with the most notable changes being

the rear window. In 1953 it changed from a small oval two-piece to a larger oval and then in 1957 this was replaced by a much wider rear window. One subtle change in 1963 was the removal of the Wolfsburg emblem that had been present since 1951.

The car had a no-nonsense, well-thought-out design. It was a very functional car, engineered to be simple to reduce the likelihood of mechanical problems. Several years after the introduction of the Volkswagen Golf, Beetle manufacture was transferred to Brazil and Mexico to take advantage of lower operating costs. The last Beetle of the classic style was assembled in 2003 in Mexico. In total 21 million Beetles were produced, making it the longest-running manufactured vehicle anywhere in the world.

fiat 500

'My mother bought "Phoebe", as she is known, back in 1969,' explains daughter Christine. 'I remember clearly the day we went to the dealership for a test drive. We were greeted by a generously vertical salesman who took pleasure in showing us around the car. When it came to the test drive he requested that the roof be open; much to my amusement he could only fit in the car with his head sticking out of the roof!

'Phoebe became a well-known sight in my home town of Scarborough, first with my mother zooming around town and then a few years later I learned to drive in her. In 1979 my mother gave me the car, which instantly made my daily trudge to work a sheer delight. However, this clocked up a significant number of miles which eventually took their toll on poor Phoebe, so she was gracefully retired ready to be restored one day.

'Whenever we have moved, Phoebe has come too – there was even a time we put her in the removal lorry! Goodness knows what our new neighbours thought when the first thing to be unloaded was a car.

'Two children later and when funds were available, we went ahead with the full restoration. At the time my mother, bless her, was still with us to see a stunningly restored Phoebe. We've since attended UK club events and have even driven to Turin, Italy, with my father as co-driver, so that we could attend the launch of the new Fiat 500.

'Putting into words my love for Phoebe is difficult...but I would say it makes me feel close to my mother.'

car notes

Post-war Italy, like other European countries, needed to mobilise the masses and the requirement was for affordability and simplicity. Fiat's solution, the Fiat 500, named after its engine size, was launched as the Nuova (new) 500 in 1957.

Designed by Dante Giacosa, it was marketed as a low-cost, economical, four-seater town car – and an alternative to the beloved Italian scooter. At only 3 metres long it redefined what was thought of as a small car. Powered by a twin-cylinder, air-cooled 479cc rear engine, with a power output of 13 brake horsepower, its instant popularity decimated the Italian scooter market, forcing many manufacturers out of business.

Fiat produced the car until 1975, with an estate variant continuing until 1977. Entire generations grew up with the Fiat 500, and such was the love it engendered it proved not only a success in its home country, becoming a Italian icon, but hugely popular throughout the rest of Europe.

30
40
50
60
MPH
LIGHTS
GEN
FUEL
OIL

morris minor

'I'm an 18-year-old who quite honestly would be much happier in the 1950s,' explains Budd. 'Quite where I have acquired my love for all things 50s is a mystery to me; music, fashion and design of that era just fascinates me. It's not a fascination my parents share – my mother loves reggae and my father loves 60s rock music, so who knows the source of the influence?

'When it came to my first car, why settle for something that would be worth a fraction of the cost it would be to insure? And the fact is, I could never really see myself driving a modern car anyway, so a classic seemed like the perfect solution, as well as giving me yet another chance to indulge myself with a 50s icon.

'Granted, a brown 1962 Morris Minor is not the first choice of car for most teenagers, but I have always had a soft spot for them. When this particular Morris Minor 1000 was seen for sale I thought I would buck the trend – after all, I believe it's all about being an individual and not following the masses. Friends soon gave my car the nickname "fudge" – and whilst trying to pass time in a traffic jam I even scratched the name on the car's dashboard.

'I'm studying politics, and such is the respect the car is given, and much to the envy of my fellow students, I'm the only one who can take advantage of the teachers' car park.'

22. my cool classic car

car notes

A British icon, this car is a much-loved classic that typifies Englishness at its best. Sir Alec Issigonis was a leading designer at the Morris Motor Company. In the late 1930s his ambition was to manufacture a car for the working classes that was high in quality, had improved performance and handling, was convenient to use but, most importantly, was affordable. Variants were available in the form of a two-door saloon and a tourer (convertible), then the iconic wood-framed Traveller estate was introduced in 1952. Early models had a two-piece windscreen and headlights either side of the grill rather than incorporated in the wing, as with later models. The curvaceous jelly-mould exterior of the Morris Minor, or 'Moggie' as it is affectionately known, was manufactured in Cowley, Oxfordshire, and had the distinction of becoming the first British car to sell a million units. Between 1948 and 1971 more than 1.3 million were manufactured.

WCO 15

austin seven 850

'Minis are in my blood; I don't feel right unless there are at least two in my drive!' Peter's fascination for Minis began as a child. 'During the summer months my family would up sticks from Norfolk and head for Italy. My father, searching for a little solitude, would set off three weeks earlier in his Mini pick-up, driving leisurely through Europe and over the Alps to Tuscany, stopping at the odd vineyard en route. Mother and I would then follow over land and sea in a Morris 1100. I would watch fascinated as, like a stork carrying a baby, our car would be winched up in a cradle onto the boat.

'My smoke-grey 1961 Austin Seven 850 was previously owned from new by a friend's father. By the early 1980s the car was looking a little tired, so he dismantled it and put all the parts in four large crates with the intention of fully restoring it one day. Unfortunately, he died suddenly before being able to do this. Having seen the Mini prior to it being dismantled, I couldn't bear to see it in that state, so in 1983 I bought it and shortly afterwards it arrived in my warehouse in the very same crates. Due to several other ongoing Mini restorations, it was five years before work began and it took a further two years to lovingly restore it.

'I compete in organised road rallies. My daughter Lizzie has been my navigator for the past three years and is doing a pretty good job – apart from one rally. We were driving through a town in Northumberland and despite the whole town's population crowding the streets shouting "left, left!" Lizzie decided to go with her better "female" judgement and insisted we should turn right – it goes without saying we ended up in a cul-de-sac and lost the rally.' (According to Lizzie and in her defence this is not the full story!)

car notes

The 1956 Suez Crisis resulted in fuel shortages and rationing in England, causing large-car sales to slump while sales of microcars spawned in numbers. Leonard Lord, head of BMC (British Motor Corporation), was not a fan of these cars and vowed to design a 'proper miniature car'. The answer was the now iconic Mini, designed by Sir Alec Issigonis. Originally, in 1959, the car was launched as the Austin Seven and Morris Mini-Minor until 1969 when 'Mini' became the sole name.

On the face of it the design appears simple but working within obvious size limitations was a challenge. However, Sir Alec created an enduring exterior and made use of every square inch, e.g. storage pockets in hollow doors made possible by sliding windows and a boot lid hinged at the bottom allowing for extra luggage. The Mini, with its many endearing qualities, was above all a highly manoeuvrable city car which was lots of fun to drive. It is one of the few cars that managed to transgress wealth – a true 'classless car', as likely to be driven by a celebrity as by your neighbour.

Many variants of the Mini were made, including the Austin Countryman, the Morris Mini Traveller and the sporty Mini Cooper and Cooper 'S', which propelled the Mini into a dominating Monte Carlo Rally winner. Although originally manufactured in England its success led to assembly plants all around the world and it was made under licence as the Italian Innocenti Mini Cooper. After a total of 5,387,862 Minis had been manufactured production stopped on 4 October 2000.

The
AUSTIN
SEVEN
DRIVER'S HANDBOOK
SHELL

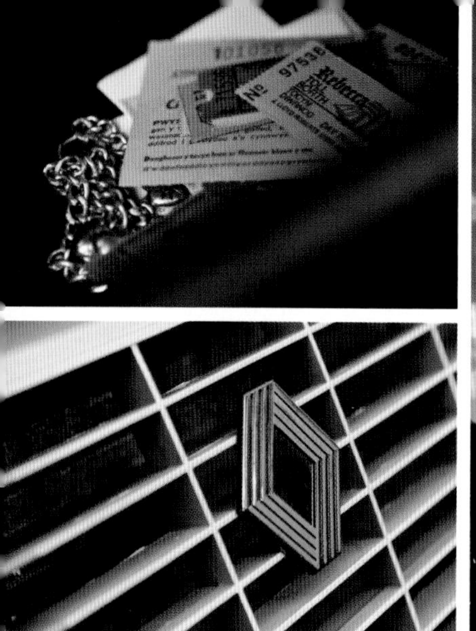

renault 4

'Whilst travelling in Morocco, I grasped the opportunity to visit the wonderful mountains surrounding Marrakech. The 4x4 I was travelling in was by no means struggling with the inclines, but to my amazement there were a steady stream of Renault 4s effortlessly making their way up the twisty roads and in some cases overtaking us,' explains Sandra. This was the start of Sandra's fascination for this understated, demure but immensely characterful French car. Sandra continues, 'When I returned home I decided to look for a Renault 4 of my own and eventually found this one, which I have named Brian. It is not a car that screams for attention, far from it, but what he lacks in sleek curves he makes up for in lovability and dependability.

'I use the car daily and, unlike other cars I have owned, I readily find myself considering what Brian is being exposed to. I try to avoid parking next to inconsiderate car owners who may dent my car when opening their doors, driving on unkempt roads or in bad weather. Brian repays this consideration by rarely letting me down.

'When we bought Brian we found an extensive collection of tickets in the ashtray from places the previous owner had visited in the car – no doubt our family outings will further add to this collection over the coming years.'

car notes

The Renault 4 (also known as the 4L, pronounced 'Quatrelle') was introduced in 1961 as a rival to the Citroën 2CV – Renault envisaged a larger, more refined city vehicle. After many years of economic struggle France saw a huge increase in car ownership as financial confidence grew.

Renault's vision paid off when less than five years after the first Renault 4 rolled off the production line they had produced a further 999,999. It was a huge commercial success for Renault – and although marketed as a small estate car it was remarkably spacious for its size. While other manufacturers were embracing adventurous exterior styling, Renault chose a deliberately restrained utilitarian design with maximum simplicity, a practical box-shaped workhorse that the French expected to last many years. The same applied to the interior, which favoured a minimalist approach with a simple dashboard, valuing comfort over show.

Over the very long 31 years of production (1961–92), apart from a few slight cosmetic enhancements, the size and shape of this classic French car have hardly varied at all.

citroen 2cv

'One selling point never mentioned in the Citroën 2CV brochure was the complementary social life it attracts. This usually centres around camping and 2CV gatherings, which brings with it an eclectic mix of fun and interesting people,' explains Chris. 'I love exploring in "Barbara", my 2CV, which has taken me from Scotland to Cornwall and all across France. Recently I went with Barbara to the biennial World Meeting of 2CV Friends in Salbris, France. Over 7,000 iconic Deux Chevaux made their way to the event, this was some sight!'

However, this 2CV, built in 1978, has not always been so loved. 'Barbara is my prized possession and is meticulously looked after – unlike the treatment it received from its previous owners. These were two students who thrashed it around Europe until bored of travel, then cast it aside in their garden. When I went for a pre-purchase inspection I found a very sorry-looking rusty car. It was neither drivable nor towable. The condition didn't faze me too much as I had the support of good friends and specialists with 2CV experience. After an extensive restoration my 2CV, in its original rare colour, "Jaune Cedrat", is now considered to be one of the best examples in Europe and to my delight has won several high-profile prizes.'

car notes

In 1948 the Citroën 2CV (*deux chevaux*) was designed to entice the French away from their horse and carts and over to a four-wheeled option. Perhaps its most famous boast is its ability to drive across a ploughed field with a basket of eggs without any breakages.

The iconic Bauhaus-inspired exterior was the work of Flaminio Bertoni, who originally trained as a sculptor. The public had a love/hate relationship with his cars as he was considered to be radical for his time. The severe but elegantly simple geometric exterior styling, along with innovative engineering, allowed for an easily maintained, economical, practical vehicle suitable

for on- and off-road driving. The adjustable ride height made it perfect for agricultural use, but above all it was amazingly cheap to purchase. Simple features included a canvas roof that opened like a sardine can and windows with hinged flap-up panes.

Over 42 years 3,872,583 2CVs were built, although variants made using identical mechanics but different exteriors (such as the Ami, Dyane, Acadiane, Mehari and Fourgonnettes), pushed the total up to nearly 9,000,000. Towards the end of its life the 2CV fell well behind in speed, safety and finesse and became unfashionable, ultimately leading to its demise in 1990.

austin 7 chummy

Maybe it was one sherry too many or a slip of the wheel, we will never know! But when Rev. Howarth of Matching Green Church left the village pub in the late 1940s in his 1925 Austin 7 Chummy, much to the consternation of the local duck community he drove straight into their pond. Maybe he was too embarrassed to be seen by villagers fishing his car out of a pond, who knows? Anyway, he left it there. A few days later he was accosted by the vicar of a neighbouring parish and asked if he intended to remove his car. Rev. Howarth remarked, 'I have another car, so if you can get it out, it's yours.' So with some help that's exactly what the other vicar did. No complicated electronics to dry out here, so several days later the Austin 7, still a little damp, was working again. In 1971, John White, the now very proud owner, acquired the vehicle. 'I remember the car well as a young lad. I was part of the church youth choir so I often saw it parked outside the church and on a few occasions was driven to Scout camp in it. Years later when I learned it was for sale I snapped it up. I have to say I was in two minds about bringing the car back to the scene of the incident,' laughs John. 'I had visions of the car desperately trying to edge away from the pond!'

car notes

Austin Motor Company produced the '7' between 1922 and 1939. Deemed the most successful car produced for the British market, it succeeded equally well abroad, quashing many other small car and cyclecar manufacturers in the process. Sir Herbert Austin felt a smaller car would be more popular than the larger cars they were known for producing and, despite objections from the board of directors, he got his way.

He invested a large amount of his own money into the design in return for a royalty for every car sold. A lower than expected 2,500 cars were made in year one – but when production ended 290,000 cars and vans had been made. Due to its success it was licensed and manufactured overseas by BMW as its first car, called the 'Dixi DA-1'. In France it was sold as the 'Rosengart LR2' and the Japanese car manufacturer Nissan mimicked the Austin 7's design for its first cars (not under licence).

The success of a re-bodied Austin 7 called the Austin 7 Swallow Saloon, produced by William Lyons, helped the foundation of what later became Jaguar cars.

morgan 4/4

'What better way to celebrate one's 40th than hiring a sports car and travelling to your favourite restaurant?' asks Tom. 'As an artist I've always appreciated the craftsmanship that goes into a Morgan. However, en route to the restaurant I was smitten by the driving experience and I just knew hiring one was never going to be enough.' A few weeks after settling into his forties, Tom went looking for his own Morgan.

Tom continues, 'I located a 4/4, a decent enough example, but I was made fully aware that a chassis-up restoration was on the cards in the near future. So before the car went to the workshop for the necessary repairs, I decided to take it on a long overdue trip to Sicily – which also gave me a chance to put brush to canvas. Despite 3,000 miles, by boat, car train and road, the car performed remarkably well and attracted huge attention wherever I stopped; children would gather round and comment on *la bella macchina rossa*.

'The bond that developed between myself and the car after our three-week trip was immense – "it" even became a "she"! It was a tough call when she went into the workshop to be restored – after all, I knew her every foible – both good and bad. Luckily the soul of the car was retained and in no time, we were nipping along Cornwall lanes. Life by the sea does take its toll on a car and she will soon need further attention, though I am happy with the way she is. In fact I love her; the car is comfortable and well used, just like a good pair of slippers.'

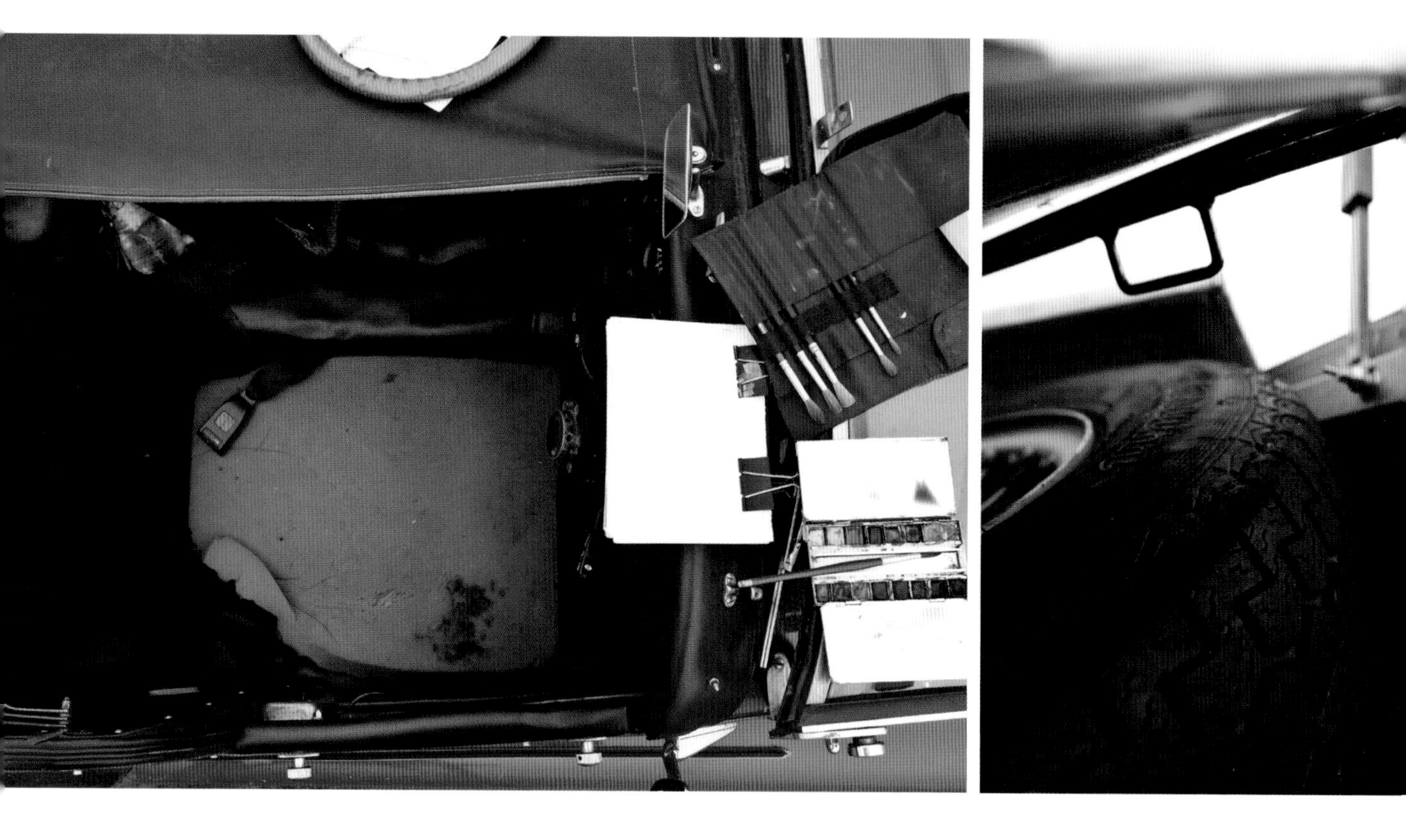

Morgan

car notes

The British motor car manufacturer Morgan was established in 1910 by Harry Frederick Stanley Morgan. When he died in 1959 his son Peter Morgan continued to run the company until his passing in 2003 and it is now in the hands of his son Charles Morgan.

Morgan has always strived for and achieved a sports car performance and experience. Early vehicles were two- or four-seater three-wheeled vehicles known as cyclecars, thus avoiding the British tax applied to cars as they were classified as motorcycles. Stiff competition from the likes of the Austin 7 forced Morgan to consider its first four-wheeled vehicle, which made its appearance in 1936. Known as the 4/4 (4 wheels, 4 cylinders) it has been in production until the present day apart from during WWII and for several years during the 1950s.

The timeless 1930s design has changed little and still features the trademark wings and running board, wire wheels and bonnet louvres, with drop-down windscreen and removable side windows for a total open-top experience. Pleasing details include the teardrop-shaped sidelights on the front wings, sculptured front and rear light clusters, exposed hinges and triple windscreen wipers.

Morgan's skilled workforce still produce their cars by hand in limited numbers using traditional techniques. The Morgan is built around an ash frame (giving it unique strength and flexibility) with a steel chassis, resulting in a lightweight sports car.

triumph herald

'My late husband, Don, had two weaknesses, cars and planes, and what he didn't know about either of them wasn't worth knowing. His love for Triumph Heralds started in the 60s, when our neighbour bought a brand new bright red Herald for their son's 18th birthday. Don loved it.' According to Iris, Don's widow, he had another incentive to fulfil his dream. In August 1985 his first grandchild was born and so the very next day Don went out and bought his beloved Triumph Herald, henceforth becoming known as a 'trendy grandad in a convertible'. With the roof down and the wind in their hair he and his two friends (all in their seventies) would drive to the golf course – what a sight...three white-haired "trendies" in a white Herald!'

Don's son, David, was given the car on his 50th birthday in 2008 and he now uses the car frequently. He plans on retirement to take part in Club Triumph's charitable Round Britain Reliability Run, where 100 Triumphs travel from London to John O'Groats, down to Land's End and back to London in 48 hours. 'It's something I know Don would be thrilled about – that his treasured possession is still well used and well loved.'

car notes

In 1959, in a bid to compete against the Mini, Morris Minor and other marques, the exciting, sleek and sporty two-door, four-seater Triumph Herald was introduced. For the first time Standard Triumph (later to become Triumph after being taken over by British Leyland in 1963), in a serious bid to compete outsourced the design of the car to Italian stylist Giovanni Michelotti – a significant figure of sports car design in the 20th century.

The result was a crisp-edged car, with later models having distinctive slanted eyebrows above the front headlights, but always retaining the small rear fins. The Triumph Herald was available in saloon, convertible, coupé, van and estate models.

By 1970 the Herald had ceased to be manufactured due to poor performance, culturally outdated styling and high build costs due to labour-intensive construction, which caused Triumph to be selling at a loss. However, over its very successful 12-year production run more than 300,000 were made.

LUCAS
SEALED BEAM
ENGLAND

retro

Retro-styled cars of this ilk are far from culturally outdated. They are instead nostalgic time capsules of wonderful if sometimes marred design – accented by styling, colours and materials all driven solely by the inspirations and fashions of the time. You may find yourself reminiscing, evoking happy (or maybe not so happy) memories from your youth, perhaps of your first car or maybe one owned by a parent or grandparent.

The Austin Allegro Vanden Plas and AMC Pacer, as featured in this chapter, having been mocked by scoffing critics for so long, are now enjoying their renaissance – becoming cool retro classics, thanks to those plucky individuals that were strong enough to wade through the doubters. The well-overdue respect now bestowed upon these worthy cars is currently reflected in their acceptance after so many years of negativity.

The preserved condition of these cars is a credit to their owners' dedication, not wanting or finding a need to change their vehicle's appearance. Also to be savoured is the charming innocence of an owner unaware that his vehicle, bought many years ago, has now become a retro classic. With this in mind you might be tempted to contemplate compiling your own list of cars to add to the 'retro car hall of fame' and consider which present-day models could become the retro classics of the future.

vw golf

'I've owned a fair few VWs over the years. Currently I'm juggling loyalty between a Scirocco Storm and a Mk2 GTI,' explains Roger. 'When a friend, a serious VW addict, added this VW Swallowtail Golf to his collection I made it my mission to prise it away from him...in an honest way, of course! It took a few months, but eventually I made him an offer which left him with a tidy profit.'

Roger often gets asked by admirers of the car what it cost to restore. 'Nothing,' is his reply, which apparently causes some confusion, until he explains that the car is 100 per cent in its original state.

'Looking back through the extensive documented history it shows the original owner, a Miss Best (very apt), did nothing more than a few hundred miles a year. It never let her down, apart from the clock, that is. The service log states ever-increasing levels of annoyance expressed by Miss Best regarding the dashboard clock that did not work. All subsequent owners have refused to repair the clock, even though I'm sure it would be easy to do, so I have carried on the tradition. I feel that if the clock were ever started it would bring about the ravages of time that many other cars of the era have succumbed to.

'The Golf will do the rounds of the VW shows and no doubt will attract plenty of attention because there are few examples of this age left...and fewer still that have avoided a teenager's tinkering.'

AA
MRX 601P

car notes

By the 1970s Volkswagen were eager to repeat the success of their much-loved Beetle, which by now had been in production for over 35 years. In 1974 they launched the first-generation Golf, a modern-styled front-engine, front-wheel-drive hatchback. This compact, economical hatchback car was designed by prolific Italian automobile designer Giorgetto Giugiaro.

This Golf, with its striking retro colour clashes and angular lines both inside and out, was a radical departure from the curves of the Beetle. Also the interior was a huge step forward, no exposed metalwork, overall a much more refined experience. The Golf was adaptable and practical and suited many market sectors – becoming something of a Euro icon.

Later Golf variants included the 1976 GTi – the father of 'hot hatches'. Such was the success that in 2007 Golf production numbers reached 25 million, making it Volkswagen's best-selling car and the third best-selling car in the world. In 1984 a bigger, wider and more comprehensive Mk2 Golf was introduced; however, for a further 25 years Volkswagen continued Mk1 production in South Africa as the low-cost Econo Golf, or CitiGolf.

rover sd1 3500 vitesse

Dennis is not a man to do things by halves. A testament to this is his perfectly manicured Japanese-inspired garden and an impressively detailed model train layout, created over the past 40 years in the 4-foot void under his house (an amazing sight). 'I'm a perfectionist,' explains Dennis, 'and I suppose that has followed through into the cars I own. I purchased the Rover SD1 3500 Vitesse new in 1983 and after a short running-in period my late wife Norma and I set off on a trip to the Isle of Man – a place close to our hearts and a long enough drive perfectly suited to get to grips with a new car. Twenty-eight years on it is still a wonderful car and holds its head up high when compared to its present-day counterparts. My compulsion to keep the car in pristine condition has been repaid with a trophy cabinet full of awards for pretty much every SD1 category. Now, apparently, the car is considered to be a "retro" classic and attracts a lot of attention at the car shows I attend – "retro" it may well be, but to me it's still the wonderful car I purchased all those years ago.'

car notes

The Rover SD1 executive car was the result of British Leyland's 'Specialist Division', hence the SD, with the 1 referring to the in-house design team's first car. The SD1 was Rover's fastback wedge-shaped replacement for the P6. In 1971 David Bache headed up the design team, finding inspiration from, among others, the Ferrari 365GTB/4 'Daytona' and a Pininfarina concept car based on an Austin 1800. The engineering was in the hands of Spen King. David and Spen were no strangers in working together after joining forces on the Range Rover. The SD1 made its debut in June 1976, with the V8 Rover 3500 Vitesse following a year later. This model included further luxuries such as a trip computer, electric mirrors and windows, and on the exterior a rear spoiler and 'Vitesse' decals. The SD1 also received the 1977 European Car of the Year award.

The success was good news for British Leyland, but the demand took them by surprise and they struggled to fulfil orders, not helped by industrial unrest. The SD1's profile rose higher through its success in international touring car racing. Production continued until 1986.

austin allegro vanden plas

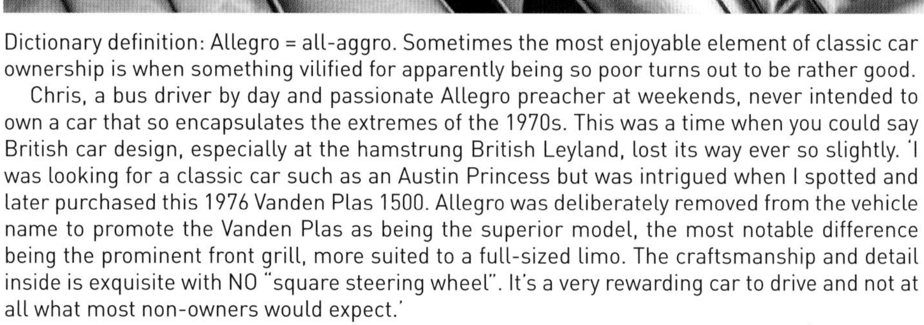

Dictionary definition: Allegro = all-aggro. Sometimes the most enjoyable element of classic car ownership is when something vilified for apparently being so poor turns out to be rather good.

Chris, a bus driver by day and passionate Allegro preacher at weekends, never intended to own a car that so encapsulates the extremes of the 1970s. This was a time when you could say British car design, especially at the hamstrung British Leyland, lost its way ever so slightly. 'I was looking for a classic car such as an Austin Princess but was intrigued when I spotted and later purchased this 1976 Vanden Plas 1500. Allegro was deliberately removed from the vehicle name to promote the Vanden Plas as being the superior model, the most notable difference being the prominent front grill, more suited to a full-sized limo. The craftsmanship and detail inside is exquisite with NO "square steering wheel". It's a very rewarding car to drive and not at all what most non-owners would expect.'

Chris now enjoys the challenge of turning Allegro sceptics into Allegro converts. 'A few years ago you couldn't give Allegros away, but recently there has been an upward trend and prices are steadily rising. I like to think that I am doing my bit to dispel some of the Allegro fallacies.'

car notes

The Allegro could be said to epitomise the troubled British Leyland of the early 1970s. The design ignored the emerging European trend towards versatile hatchbacks and instead wrongly stuck with a booted option. Furthermore, rather than adopting the trend of sharp wedge-shaped exteriors they settled with a more rounded body, partly to accommodate existing engine parts and thus keep costs down.

However, with the Allegro's upmarket sibling the Vanden Plas the purse strings were loosened. A new front grill; reclining leather seats; fold-down rear picnic shelves; deep luxurious carpets; walnut dashboard and a host of other improvements. This all came at a price, specifically a sale price in 1974 of £1951 over £1159 for the basic Allegro.

While it is true that the Allegro had problems, they were nothing like on the scale that the press reported. Fallacies such as 'that's the car with the square steering wheel!' (it was 'quartic', never caught on and only appeared on the car for a very short period of time), have given it an unjustifiably bad press. Allegro sales were strong over its 10-year life, though, and sales totalled 642,350. It was succeeded by the famous 'talking' Maestro in 1983.

hg holden ute

'My first automobile was a well-worn 1949 MG TC Roadster. Disconcertingly, its brakes failed during the drive home, requiring a quick application of the handbrake to avoid a crash!' explains Ken. 'I then made a rash decision to fully restore the car. So, I carefully stripped every nut and bolt down to the bare chassis, then rebuilt it and painted the panels in British Racing Green. My labour of love took 12 months, but the reward was worth it. I no longer had to borrow my father's FB Holden – or worse, my brother's clunky Austin A40. Eventually I traded the Roadster for a devilish Triumph TR4 – fast, reliable, with modern conveniences like a radio and wind-up windows. Then came a variety of old English cars that somehow just kept going. One epic journey I made was 8,000 kilometres from Sydney to Perth in a 1967 Hillman Imp GT – a small car with a big heart. When I spotted this 1971 Holden HG utility sitting forlornly in a friend's paddock I couldn't leave it there and just had to have it! I enjoyed restoring it back to a classic look with a new paint job, tonneau cover and renovated seats. I didn't want it in perfect concourse condition because owning a ute makes you everyone's best mate – and I would never agree to move half the things that I do.'

car notes

An Australian 'ute' (utility vehicle) is considered among purists (and to honour ute designer Lewis Brandt's vision) to be typically based on a passenger sedan, for instance a Holden HG four-door saloon. The design was modified in production so that it could accommodate an integral cargo bed with a continuous seamless body.

The ute design was based on the plea of a farmer's wife in 1932 to Ford Australia, to build a car that could take her to church on a Sunday and carry her pigs to market on a Monday and enable her to stay clean in the process. The plea was acknowledged and Lewis Brandt's concept of a dual-purpose farmer's or tradesman's vehicle rolled off the production line in 1934. This icon of Australian life is very much still in existence despite many unsuccessful overseas imports attempting to replicate it.

amc pacer

When the AMC Pacer was introduced in the mid-1970s, it was always destined to be a risk with its grandiose futuristic styling. Traditionally you were either a follower of Ford or General Motors, and AMC sat, often overlooked, somewhere in the middle. During the Pacer's development, several design issues saw the idea of a compact, economical car (to combat the 70s high fuel prices) turn into something of a difficult-to-categorise vehicle, compact measured against other American cars on offer, but still large compared to its European counterparts.

Becky, the rightfully proud owner of this quirky vehicle, continues: 'The Pacer was bought for me by my husband, Jamie. He has many other classics including an AMC Gremlin and a Ford Country Squire complete with simulated vinyl-wood sides. I don't know why he enjoys owning such "cool" oddities from motoring history, maybe it's a result of not wanting to follow the masses and a craving to own something different. It certainly attracts attention wherever we go! I love driving it and judging by the increasing numbers of loyal followers I'm not the only one, as it's now being regarded as a cult classic.

'We live near to several American Air Force bases and Jamie has befriended a few service personnel. Some have now relocated back to America and have become his scouts, informing him of anything interesting they find. Who knows what else he will obtain in his collection.'

CAFE ~ DINER
Wall's

1,25 litre
Coca-Cola
Coke
Budweiser
LAM
436P
LAM
436P
12
6
MAX COLD
OFF
A/C
HEAT
HI
FAN

car notes

During the early 1970s many American manufacturers who hadn't predicted the steadily increasing petrol prices, due to the oil crisis, had little to offer the consumer from their gas-guzzling ranges. In 1971 AMC (American Motors Corporation) showed foresight when its chief stylist Richard Teague started to conceive a more fuel-efficient compact car as well as something fresh, unique and futuristic.

In an era of slab-sided automobiles the Pacer was quite unlike anything else when it was launched in 1975. Although compact by comparison, the Pacer was the same width as a full-sized American car, giving occupants the impression of being in a larger car, as shown in AMC's advertising: 'the first wide small car'. Another unique element was the lengthened passenger door to assist access, while the large glass area led it to be dubbed, among other names, the Flying Fishbowl.

During the first two years AMC's risky strategy seemed to be paying off, but then safety modifications increased vehicle weight and thus lowered performance and economy. Sales were further hit by the influx of smaller imported cars. After building 280,000 cars, Pacer production ended on 3 December 1979.

rover p6

'In 1980, while my friends were souping up their Ford Escorts with fake twin exhaust pipes, huge speakers and the obligatory go-faster stripes, I was hankering after a choice of car that, at the time, was perceived as a "grown-ups" car,' explains Barrie. 'Owning a car like a sedate Mexico-brown Rover P6 did make me the odd one out, which I rather enjoyed. As a teenager my father and I would spend our spare time tinkering on cars, which is most likely why I had a thing for 70s cars; I understood the mechanics and their quirky design ideas, which gave me an added confidence when it came to buying one.

'My original Rover most likely ended up seeing out its final days in a scrap yard, but I hadn't quite got the P6 bug out of my system. In 1983 I ended up buying this 1976 V8 3500 S in a rather fetching shade of avocado green (a colour synonymous with the era). You could say I have grown up in a car which now befits a man, like me, of advancing years. The car is cherished and spends the harsh winter months safely tucked away in my garage.'

car notes

The Rover P6 saloon car was available from 1963 through to 1977 and defined comfort and style for the 1960s with its elegant features. Initially the car was marketed as the Rover 2000 (denoting its engine size) and was a completely new ground-up design, aiming for mass appeal with its stately, streamlined body with lengthy curves disappearing to narrow fins at each corner. It was considered ahead of its time for its safety, advanced suspension and transmission. A novel idea was the prism of glass on the top of the front sidelights to signify the corners of the car in poor light. All these innovations limited luggage space, so if required the tyre could be relocated to the boot lid. In 1968, the same year that Rover was brought under the British Leyland umbrella, they introduced the 3.5-litre V8 engine. This increased the top speed and 0–60 time to an impressive 114mph and 10.5sec respectively. Despite the P6's size the interior space was not generous and the rear-seat design had room for only two.

As the design started to date British Leyland were well under way with the P6's replacement, the SD1. The last P6 was built on 19 March 1977 – but not before 322,302 had been produced.

GT

toyota celica gt2000

'I've always been into modifying things,' says Duncan. 'At the age of 12 my brother and I had a Raleigh Runabout, basically a beefed-up pushbike with an engine strapped to it, and to make it more like a motorbike I shifted the petrol tank from the rear to the front. This is the sixth Celica I've owned. The first, which was kindly given to me, had a matt black exterior with a cream vinyl interior. Totally ratty-looking with an exhaust you could hear half a mile away – you'd never get away with it nowadays. My second was a MkII RA40 Liftback. The third and fourth were taken apart and built into what effectively was my fifth.

'Whilst collecting some spare parts I was shown the vendor's car tucked away at the back of his heated garage. Instantly I knew I wanted his 1977 Toyota Celica GT2000 Liftback but didn't have the money, and anyway he didn't want to sell it. Several years later I had a telephone call from him – he'd changed his mind and asked if I was still interested? I was, and after raiding my savings account I phoned him back. I brought her home in 1998 and right away knew I had one of the best examples in the country. Soon after, I entered her at JAE (Japanese Auto Extravaganza) and came second, so from then on she was known as "Second Prize".

'Thankfully, I'm a lot more sensible now than I was in my youth and have learned the error of my ways, but the car still gives me a Peter Pan-like feeling – which is something I don't want to lose. Next year we're off on a very overdue trip to the Spa Classic – which will most certainly blow away any cobwebs!'

car notes

In October 1970 Toyota introduced the Celica range of cars to the Japanese market at the Tokyo Motor Show. It was Toyota's aim for the Celica to be a more affordable alternative to its top-end 2000GT sports car and they promoted it as a car that highlighted style and maximised the enjoyment of driving. Engine capacities were kept to a modest size (under 2 litres) to conform with Japanese regulations.

In 1975 the Celica was given a facelift with a revised front bumper and grill layout – overall a more Americanised style was introduced. The Liftback was available to the Japanese market in 1973, but overseas buyers had to wait until 1976. The Liftback was often referred to as the 'Japanese Mustang' because of the apparent similarities to the Ford Mustang by way of the fastback profile, side and bonnet air intakes and embossed interior logo, along with other styling influences derived from the muscle-car era.

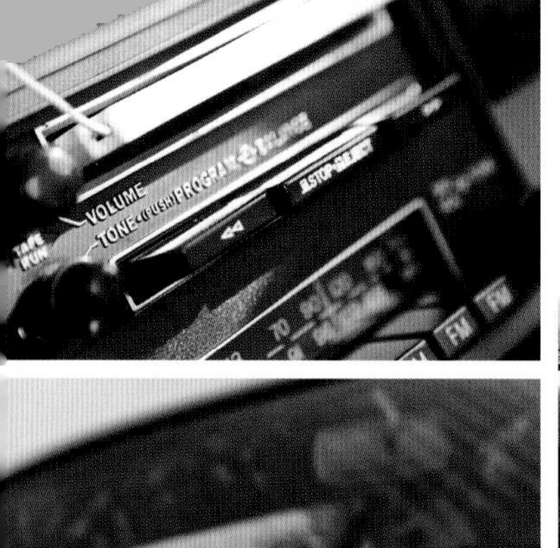

GT
2000
RMX
252R
DEFOG
TURN

ford capri mk1

'My first manifestation of mobile independence was a two-wheeled pedal racing bike. This didn't hang around for long when my father, a police officer, realised it was made up of a number of stolen parts,' explains Ed. 'In the 60s I found rock'n'roll, to my father's dismay. I spent a whole week's paper-round money on the Rolling Stones' first album, which my father thought was a total waste of money – respectfully, he was wrong, as to me it was the start of an obsession. During the mods and rockers era a scooter was the "must have" accessory, but annoyingly I kept falling off.

'I persevered, but with my balance being so poor an additional wheel seemed like a good idea, so then came the acquisition of a British Racing Green three-wheeled Reliant Regal van with a hand-painted orange interior. It was only when I moved over to four wheels that I felt at home and, to be more specific, only those made by Ford. I bought, used and sold a succession of Fords until I purchased my first Capri – I was smitten. I then made it my goal to own what I have now, a 1972 Mk1 1600 GT XLR. Owning the car led me to meet my wife Karen, who organises a local classic car show. She too now owns a Capri.

'I know most car owners won't admit to this, but it's a great feeling when you catch a sneaky peek of your reflection whilst driving past a shop window, it makes me feel so proud. It's mine, it's my gem, my very own diamond. Like all the Rolling Stones memorabilia I own, the car is a good investment and it is something to fall back on if times are hard – but, trust me, the car would be one of the last things to go.'

DPN 710K

car notes

In the late 1960s Ford resolved to reproduce the success of the North American Ford Mustang and make available a European 'pony' car – compact, affordable and sporty. Responsibility for the styling was undertaken by Colin Neale at Ford Dagenham.

Production began in November 1968 and the car was launched at the Brussels Motor Show on 24 January 1969. Twelve days later a Ford Capri was available to view on every dealer forecourt. Ford was right – the European longing for a slice of the Mustang Pie made the Capri a great success, with 400,000 cars sold in the first two years. Much like its American cousin, the Capri Mk1 success story was in part due to its affordability.

The Capri bore many similarities to the Fastback Mustang with the long front and bonnet bulge (available on certain Mk1s), vinyl roof, side air intakes and a long swooping side crease. Capri interiors, on certain models, also bore hallmarks of Mustang inspiration with a sporty three-spoke steering wheel, heavy use of ribbed vinyl upholstery and wood fascias.

The Capri was sold as 'the car you always promised yourself' and over the three revisions – Mk1 (1969–74), Mk2 (1974–8) and Mk3 (1978-86) – and 18 years of production, two million people did just that.

'Like most teenage boys, my bedroom walls were adorned with various posters of the era,' explains Enzo. 'Amongst the typical iconic Italian Supercar posters was a Fiat X1/9 Bertone – to me this was automotive perfection – an affordable, compact mid-engine sports car – and at the age of 16 I vowed that one day I would have one of my own. Unfortunately, by the time I was in a position to buy one, good examples were in short supply, having succumbed to the automotive devil that is rust – a problem that plagued many Italian cars of that era.

'Eventually this X1/9 Gran Finale came up for sale. It had somehow survived the rust and was in pristine condition; I needed no convincing and purchased it. Sometimes the reality of finally possessing something that you have longed to own for years can be a disappointment – but not in the case of the Fiat, it lived up to all my teenage expectations and then exceeded them,' remarks Enzo.

'I don't do a big mileage in the car, maybe 500 miles a year, but those I do I love with a passion – that's the Italian in me. My young children Vincenzo and Chiara have already taken a shine to the car and as family is so important to me it will most likely be handed down to them, so that they too can experience the same exhilaration as I do.'

fiat x 1/9

car notes

At the Turin Motor Show in 1969 Fiat unveiled a concept car named the Autobianchi A112 Runabout. The distinctive wedge-shaped appearance was styled by Marcello Gandini of Bertone and took inspiration from contemporary powerboat designs.

Several years later the Autobianchi served as the basis for Fiat's two-seater, mid-engined, production sports car, the X1/9. It retained a large proportion of the concept car's aesthetics, along with the additional lightweight removable hardtop, air intakes and pop-up headlights. Manufactured by Fiat from 1972 until 1982, production was taken over by Bertone and the car rebadged the Bertone X1/9 until production ceased in 1989. The final production models, named the Gran Finale, are identifiable by the additional rear spoiler, Gran Finale livery and a liberal use of retro-patterned Alcantara fabric on the interior.

The rather prosaically named X1/9 takes its designation from Fiat's prototype codes, marking it out as the 9th passenger (X1) vehicle they developed.

panhard

panhard 24ct

'The saying "the early bird catches the worm" is true, but it took me three years to finally get it!' explains John. 'In 1978 I was working as a milkman and whilst doing my early morning rounds I would pass by a car poking out of a farmer's garage – it had been there for ages. Being a fan of French cars I knew immediately what it was – a Grasshopper Green 1964 Panhard 24CT. I felt this amazing car was just wasting away, so whenever I saw the owner I would pester him to sell it to me – which he would always flatly refuse to do.'

After three years of asking the farmer John had very nearly given up, until one day instead of his usual reply he got a 'maybe'. John continues, 'I was somewhat taken aback; the owner explained that he was emigrating and needed to sell it. So finally in 1981 she became mine.'

This wasn't the first obscure French car John had owned – he has a track record of them, including a Citroën 2CV4, Citroën Ami 8 and a Citroën AK400 van. In the 1970s he even owned a Isetta 300 Bubble car that he used as his everyday run-around.

'My wife June owns a retro Ansfold caravan and we use the Panhard to tow it when travelling to rallies.'

If only I had met John and June a few years ago I would love to have included June's caravan in my first book, *my cool caravan*. *C'est la vie!*

car notes

In 1890 Panhard was one of the world's first mass automobile manufacturers. In subsequent years Citroën became a 25 per cent shareholder in Panhard. Both French marques worked together producing a formidable range of small, medium and large cars to plug any gaps in their respective ranges.

On 23 June 1964 the two-door, four-seater Panhard CT (24 after the Le Mans 24-hour race in which Panhard were involved, and CT standing for *Coupé Touristique*) was shown to the press. The CT sported a striking and futuristic low sleek body made possible by the ingenious chassis that, without structural weakness, allowed for narrow window pillars, giving good visibility and an airy feel to the well-appointed interior. The now familiar aerodynamic front end and twin glass-covered headlights that is now recognised as the face of the Citroën DS was in fact developed first for the Panhard. An air-cooled, two-cylinder 850cc engine that first appeared in 1952 in the Panhard Dyna was selected for the 24 CT.

The anticipated high sales were not achieved and in 1965 Citroën took over Panhard. Capacity-starved Citroën then had to make the unenviable choice between continuing with the Panhard 24 development or giving over capacity to the popular 2CV vans. The vans won the toss and on 20 July 1967 production officially ended.

CHAUFFAGE

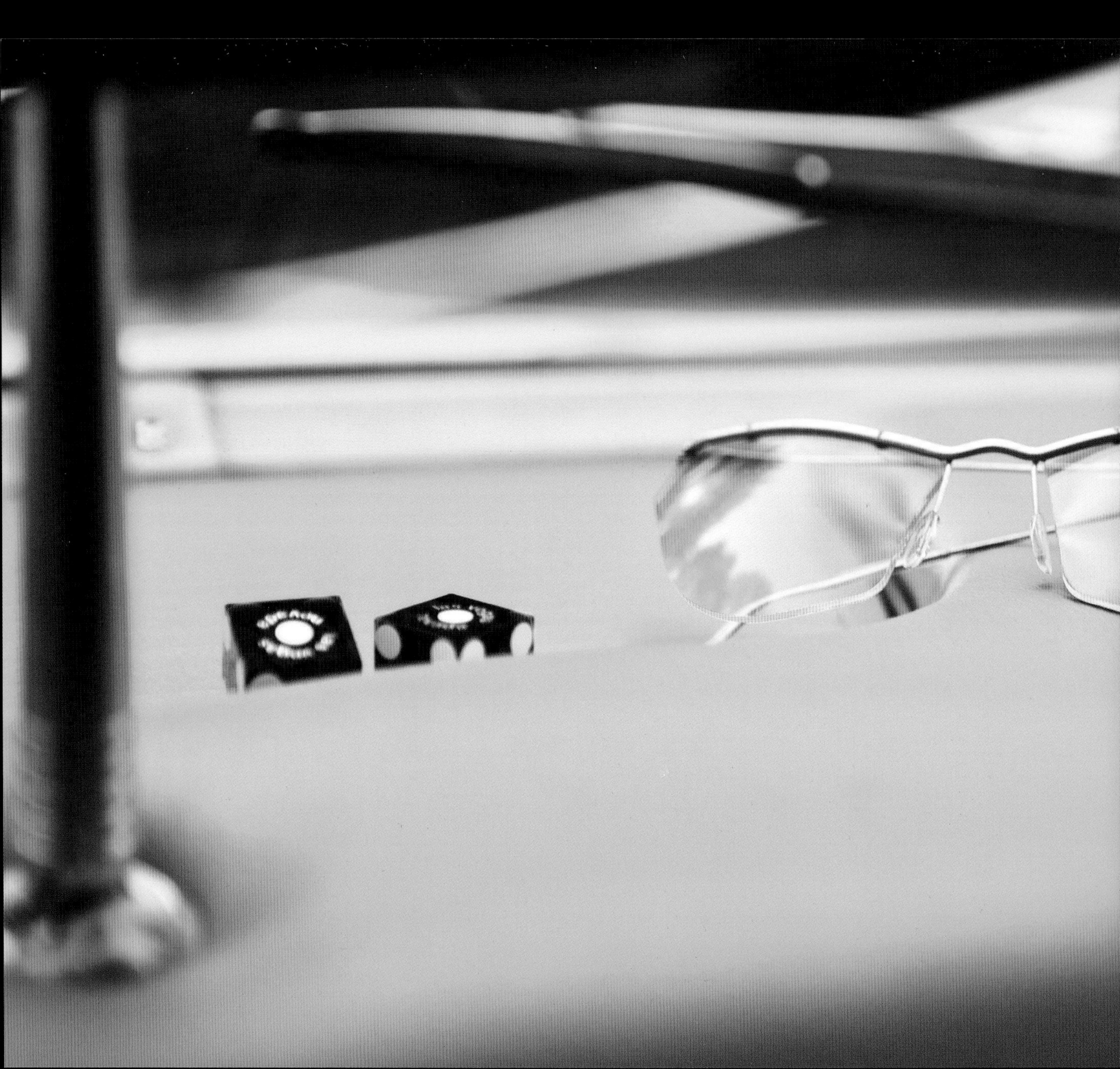

glory days

Despite the advantages of modern motoring, with its air-conditioned, computer-controlled reliability, who wouldn't want to return to those fast-paced times when automotive design was in its prime? The diverse cars in this chapter, shown in chronological order, are just a snapshot of deliberate extremes in car design around the globe that speak volumes for the can-do spirit of the age.

There was, for instance, the restrained approach of the Bentley R Type Continental with all its effortless and magnificent grandeur – so graceful in motion, with no need to shout its arrival. Then, by complete contrast, look at the bold, daring and flamboyant Plymouth Belvedere; without doubt, the epitome of wonderful excessive design. It appeared the consumer was being force-fed designs that were neither practical nor economical, with designers seemingly being given a free hand to come up with the most hedonistic creations to ever hit the highway, striving to out-design rival marques with more outlandish creations.

British racing success was translated to the road in the form of the Jaguar E-type. The Facel Vega, so keenly adopted by stars, courted notoriety for its marriage of American V8 power and exquisite French flair. Then came the Mustang, considered the pioneer and cornerstone of many muscle-car imitations.

With those thrilling glory days well behind us, perhaps never to return, it's left to the owners of such cars to celebrate and relive the experience, and for the rest of us to gaze on in admiration.

bentley r type continental

'Frank Dale started trading in 1946 to fulfil the demand for quality cars in post-war London and over the years Frank Dale & Stepsons (myself and brother John being the stepsons), have handled some of the best-known examples of these wonderful motor cars. In 2011 we celebrated 65 years of trading, making us, to my knowledge, the oldest independent Rolls-Royce and Bentley specialist in the world,' says Ivor. 'In 1954 this elegant Bentley R Type Continental was attracting the crowds at the Geneva motor show. The car was then bought by an entrepreneur and remained in Switzerland for some years; it subsequently changed hands and went to the south of France. It eventually became mine in 1972. Despite the car's rarity I use it regularly – whisking me to work in blissful comfort along the M4 corridor. From time to time we make longer journeys, to the south of France and Italy.'

'In or around 1990 Dinky Toys requested that my cherished Bentley be made into a 1/43rd scale model toy, so my car was measured from front to back in every detail to create an exact replica – even down to the number plate! It's not something you get asked everyday, it was a real honour.'

car notes

For those that could afford it fewer things could be as rewarding than when purchasing a Bentley R Type Continental. In 1952 it was the second in the series of post-war Bentleys and the first appearance of the Continental name. It was then, and remains today, an automotive work of art. British craftsmanship was in full flow as the post-war rationing that only allowed individuality to go so far began to come to an end (petrol rationing ended in 1950). The streamlined body and fastback shape with curvaceous, slightly finned rear wings (for directional stability) concealed an impressively fast car – at the time, it was quoted as the fastest four-seater production sports car in the world.

A rolling chassis was delivered to your choice of coach-builder, of which H. J. Mulliner & Co. was the preferred choice. Once there your individual requests, both inside and out, would be adhered to, including opulent and extremely well-refined interiors on a par with a bespoke Savile Row suit. Exclusivity was further enhanced by just 208 R Type Continentals being built, of which 193 were the work of H. J. Mulliner.

corvette v8 speedster 'duntov'

'The Corvette Duntov EX-87 is a favourite of mine,' says Tom, 'as it's a one-off; ownership was not an option. It's probably now idling away as a centrepiece in a private collection, unlikely ever to see the road again. My appreciation for the shapely lines of Corvettes began and developed at university whilst studying architecture. I acquired my first Corvette, a 1966 327/350 convertible, from Eric Burdon of the pop group The Animals. I am extremely lucky that my passion for Corvettes has developed into a business, having bought and sold over 700 of these amazing cars since the early 1970s. To fulfil a long-held dream I turned my hand to replicating a Duntov EX-87. A suitable donor car was at my disposal so there was nothing to stop me. Once completed, and in recognition of all my hard work, I drove it to the 2011 Le Mans 24-hour endurance race – a stunning drive – and, as an added bonus, the car was chosen by the ACO to lead the *Parade des Pilotes*.

'Surprisingly I travel to work not in a car but on a pedal bike – I'm opposed to the nonsensical trend of using a car unnecessarily for short distances. As someone that truly embraces the experience of driving I feel it should be savoured and not misused.'

car notes

The Chevrolet Corvette was first shown to the public in 1953. The rounded exterior was designed by Harley Earl, with its mesh stone-guards over the headlights, toothy front grill and tail-lamp fins all enhancing its sporty appearance. Much-needed improvements were made when Mauri Rose, the Chevrolet consultant and former three-time Indy 500 winner, suggested that a preproduction version of Chevrolet's 1955 V8 engine should be installed into a 1954 six-cylinder Corvette. This was done by Zora Arkus-Duntov, the talented Chevrolet engineer and Le Mans driver, who changed the car into a high-performance prototype. His vision was all about pushing the Corvette to the forefront of sports cars, a match for its European rivals.

Further modifications were made: a lightweight Plexiglas windscreen, passenger side tonneau cover and, in imitation of racing Jaguars, a glass-fibre headrest/tail-fin to assist with high-speed stability. The EX-87 is an important part of the Corvette's history and the starting point for all the great Corvettes that have followed in its tracks, with Arkus-Duntov considered a legend among Corvette aficionados.

plymouth belvedere

'Some say that an 18-foot, 1.5-ton classic car with a turning circle like an ocean-going liner is a brave choice for your first classic car. My view is that there is little point settling for something you're just not happy with,' explains Paul. 'Myself and Patsy, my wife, enjoy jive dancing and the 50s/60s rock'n'roll vibe, so the car fits in just nicely with our lifestyle. However, it can't be said that the car fits in anywhere else – I do try to be considerate when parking but inevitably take up two spaces.

'I looked at a number of American cars before I bought the 1960 Plymouth Belvedere, but the flamboyant styling and aviation-inspired, high-flying fins generate something amazing to see from every angle – so there was no other car for me.

'Even inside it doesn't disappoint, with its quirky features like the square aerodeck wheel and push-button transmission – it was almost as if the designers were using up all their ideas in one go in case an atomic nightmare unfolded.

'For over 50 years I managed to resist the temptation of a tattoo, however the car has wormed itself into my life in such a big way that I am now adorned with a tattoo of the car on my forearm,' Paul confesses.

'To assist the running costs (5.2 litres of American V8 doesn't give much of an mpg return) I make the car available for the occasional wedding. It brings about smiles from both the bride and groom and the congregation as we effortlessly glide away from the church – it also gives me another excuse to put on my blue suede shoes.'

car notes

In 1951 the first-generation Plymouth Belvedere two-door, hard-top coupé was unveiled. Initially it was not a separate model but instead the top-level option for the Cranbrook range of cars and called the Cranbrook Belvedere. In 1954 the second-generation Belvedere became a model in its own right, superseding the Cranbrook range. It was available as a convertible, two-door station wagon, four-door sedan or two-door sport coupé, incorporating styling updates like the first signs of small chrome tail-fins on the rear fenders.

In 1955 Chrysler stylist Virgil Exner made big changes to the range and once again rightfully returned the Belvedere to the top spot with his 'forward look' inspiration. These were most notable in 1956 when further changes included a pair of rakish tail-fins with chrome embellishments; it was also the first ever American car to feature a push-button, automatic transmission. In 1957 Chrysler sought to highlight the ahead-of-its-time Belvedere with the slogan 'Suddenly, it's 1960!' Further developments continued with quad headlights and a big block V8 engine dubbed the 'Golden Commando'. This Belvedere was given a signature 'saddleback' two-tone colour treatment.

In 1962, most experts would agree, the Belvedere was hit with the 'ugly stick' in a sudden departure from its high-flying design to a more common-sense approach – the end of the glory days were in sight.

jaguar e-type

The 70mph speed limit on British motorways can partially be attributed to the E-type, after a few foolhardy types tested the patience of the police on the newly opened M1 motorway. However, you cannot hold that against a car that, after 50 years, still turns heads.

'In 1961 when the E-type was launched, I was 16,' recollects David, owner of this 1973 series 3 V12. 'I thought it was the most beautiful car I had ever seen and longed to own one. In 1988 I had accumulated a bit of capital and at the time the stock market had suffered a severe downwards realignment, so I convinced myself that buying a classic car was a good investment. It was, but only if I had sold it within a year of purchase...which of course I didn't. I have a PhD in hindsight!

'I can't say the car has been cheap to maintain, a £5,000 engine rebuild holds testament to that! Again, with hindsight, I was a bit rash buying a car from an advert in the Telegraph newspaper – a little more research would have taught me what to look out for. It was solid enough but a bit tatty and parts are expensive, but as the saying goes "love is never cheap".

'The E-type has an insatiable appetite for fuel. On one occasion whilst being driven home in the cab of a swelteringly hot low-loader, after the E-type had a mechanical hiccup, I made a futile attempt to lighten the mood by remarking to my wife, "Think about all the fuel we're saving". It didn't go down well.

'The VT in the car's registration, 1892 VT, stands for Vee Twelve. However, in the 23 years of ownership not one person has ever made the connection. A bit too subtle I think!'

car notes

When 1960s Britain was finding its feet again after a turbulent period, the Jaguar E-type took a massive stride to increase the population's confidence and wave the flag for 'Cool Britannia'. The car couldn't fail with its breathtaking looks and inherited race-car pedigree, its racing heritage being taken from the Jaguar D-type Le Mans legend.

When launched at the 1961 Geneva Motor Show the E-type was a show stopper. Gone was the outdated look of the XK series and in rolled the sleek teardrop aesthetics of the E-type. Available as an open-top or coupé side-hung hatchback,

the plush interior was true craftsmanship, with careful detailing worthy of a piece of priceless furniture. And the E-type's looks were not concealing a shortfall in performance, far from it, as a slightly modified press car reached the magical speed of 150mph.

A surprising aspect of the E-type was the price, as suddenly other prestige marques such as Ferrari seemed somewhat overpriced when an E-type was available for a third of the price. By 1975 a production run of 70,000 was evidence of the Jaguar E-type's huge success.

GB
432 EMW

facel vega

'For nine months of the year I travel the world commentating for Formula One.' This is the enviable routine for Bob Constanduros, journalist and the 'voice of motorsport', who recently celebrated his 500th race commentary with, no doubt, one of his trademark elongated podium cries of 'Champaaaaagne!'

'My career in motorsport is the result of a long-held dream that started in the late 1970s, when I packed my bags and headed off overseas in my campervan in search of a career in F1...but that's another story for another time! When returning home for a few days between races, my Facel II is a wonderful sight to be welcomed with – second, of course, to my wife!

Facel first came to Bob's attention in the 1950s when he spotted an HK500 parked outside his local church. 'I marvelled at the spectacular interior and the aggressive lines of the bodywork. My 1962 Facel II was originally owned by former work colleague Rob Walker, the racing patron of Sir Stirling Moss in the 1950s and 60s and a wonderful character. His passport simply stated his occupation as "Gentleman". Rob owned the car for four years, during which time he had the speedometer re-calibrated so that Betty, his wife, thought they were going slower than they actually were. Years later it resurfaced and I purchased it from a Facel specialist in 1984; it was in a sorry state and over the next 12 years it was restored. During this time I decided to buy a roadworthy V8-powered Facellia convertible. The only problem now was that I needed something for garden parties, hunt balls and film premieres, so I bought an Excellence, Facel's suicide-door limousine. As you can see, some 50 years after I first saw a Facel I'm still completely captivated by them.'

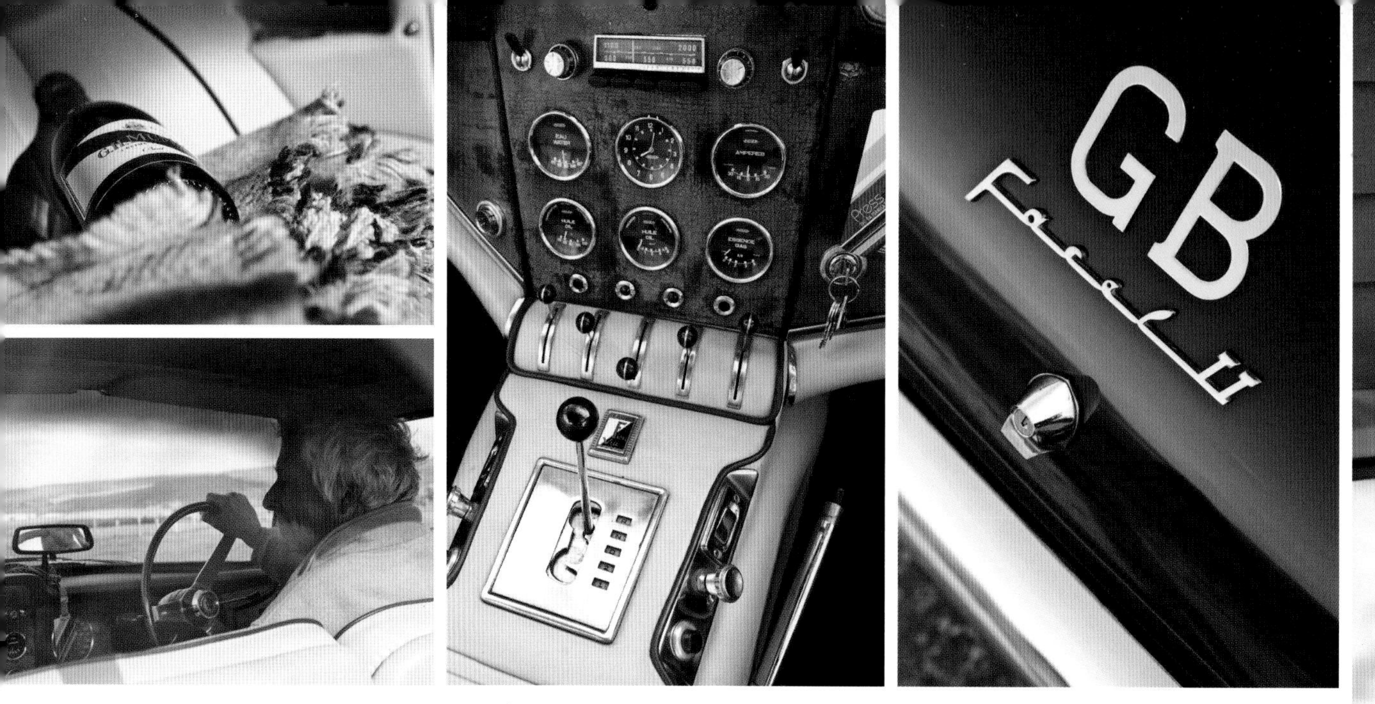

car notes

Jean Daninos' first production cars (the Facel FV and later the HK500) appeared in 1954 when he established the Facel Vega marque, introduced at the Paris Auto Show along with the marketing slogan 'For the Few Who Own the Finest'. The straightforward approach to engineering led the cars to be heavy, but performance was brisk when powered by the Chrysler V8 engine.

Despite initial early successes the 1960s saw the company face bankruptcy and Jean viewed the Facel II as his last-ditch attempt to create the ultimate French grand touring car. The car was extremely exclusive (approximately 180 were built) with a price tag to match. The chic exterior with its almost wraparound expanse of glass and elegant aviation-inspired interior was a sheer delight. In turn these stunning looks attracted many famous owners, such as Pablo Picasso, Tony Curtis, Christian Dior, Ava Gardner, Frank Sinatra, Ringo Starr, Stirling Moss and Rob Walker. The car was advertised as the fastest four-seater coupe in the world. In certain configurations the car could out-perform the Aston Martin DB4, Ferrari 250GT and Mercedes-Benz 300SL in a 0–60 sprint – all of which were two-seater cars.

Despite all these positives, in 1964 the Facel company went into receivership and production ceased. When the company closed Jean Daninos observed, 'The HK500 was the most interesting car we ever made but the Facel II was by far the best. It was totally "elegant".'

ford mustang

'At the last count I have owned 72 cars and probably driven over one million miles. My first car was a Standard Avon bought for £20. Since then I have owned a Bristol, an Aston Martin, BMWs, Volvos and many more which are documented in my own photographic log of cars,' reminisces Martin. 'The first Mustang I owned was actually a result of a misidentification. In 1989 whilst passing a car showroom I spotted what looked like a white Triumph Stag. I pulled in for a closer inspection, which revealed that it was in fact a 1970 convertible Mustang. Since the car had been on the forecourt longer than the salesman would have liked I impulsively made him a silly offer, half-expecting a terse reply of course. Surprisingly it was accepted – with the proviso that I took it away that day. Driving away in my new Mustang I found myself with the dilemma of "how do I explain this to my wife?"

'Fast-forward to the present day; having owned ten Mustangs, including a sought-after 1967 Shelby, I currently own this stunning gold 1966 Mustang Coupé. I like to think I would have owned a Mustang regardless of whether my offer was accepted or not all those years ago – because to miss the experience of owning and driving a Mustang, such an important car in automotive history, would have been a crying shame.'

car notes

The Mustang is one of Ford's most well-known vehicles and it took America by storm on 17 April 1964, assisted by an intense newspaper and television campaign with the advertising slogan 'the car to be designed by you'. On day one four million people visited showrooms with first-day sales reaching over 22,000 and record-breaking first-year sales of over 417,000.

The genius of Lee Iacocca's (vice president and general manager of Ford's car and truck division) vision for the Mustang was that it was affordable for everyone and, with four seats, family-friendly too. As the Mustang's popularity grew so did the options available, including the now sought-after 'Interior Decor Group', more commonly known as the 'Pony Interior' (embossed running ponies on the seat fronts) alongside the pistol-style interior door handles and an overall 'cowboy' styling.

The 'pony class' American car – sporty stylish coupés with long bonnet and short rear boot – was a result of the Mustang concept. It proved an inspiration for many other manufacturers in the US and overseas. However, the Mustang is the only 'pony' car to have remained in production over four decades.

EGW
297T
HERCULES

pontiac firebird trans am

'I cannot deny that with my choice of car I may be trying to emulate a certain (moustached) 70s film star, as I did when I owned my first Pontiac Firebird Trans Am when I was 23,' admits Steve. 'I probably developed my love for big engines when I started working as an engineer for Aston Martin. I then switched career and decided to join the police force – making the Firebird an ironic choice of car for a police officer, considering how it is often portrayed in the movies. Sadly, after seven glorious years I had to sell my trusty sidekick due to family commitments.

'Having been a serving police officer for 20 years, and after one particularly tough year, my wife told me to indulge myself and buy another Trans Am. I already knew of a specialist dealer who had a stunning 1979 Firebird Trans Am for sale – the last of the big-engine years. I struck a deal and then counted down the days until I picked it up. There is nothing shy about this car – from the smooth 6.6-litre V8 engine and the Firebird decal on the hood through to the gold detailing on the bodywork and interior.'

For many, a 6.6-litre engine would be more than adequate, but plans are afoot to install an Oldsmobile 7.5-litre version – as Steve openly admits, he just can't help himself!

car notes

A 'muscle car' is simply the practice of putting a big engine in a car for increased straight-line speed, often eschewing sophisticated engineering or European styling – the Pontiac Firebird epitomises such a car.

First-generation Firebirds (1967–9) featured the 'Coke bottle' body styling and were manufactured by the Pontiac division of General Motors. Second-generation Firebirds (1970–81) took on a more 'swoopy' aerodynamic shape, helped by the colour-coded moulded front bumper and grill, giving it a bumperless appearance. The Trans Am was a performance package for the Firebird consisting of uprated suspension, improved handling and increased horsepower, along with visual changes in the form of a firebird emblem that covered the Trans Am bonnet, speciality spoilers, fog lights and wheels.

The Firebird was well on its way to becoming a legend – thanks in part to its starring role in the *Smokey and the Bandit* films with Burt Reynolds – and continued on to a third and fourth generation before ceasing production in 2002.

EGW
297T

classic eccentrics

From a well-used Model T Ford that has seen better days, to an iconic DeLorean known the world over for its forward-thinking design and subsequent movie stardom, this chapter is all about stepping away from the ordinary and embracing individuality through eccentricity.

Owners of such idiosyncratic vehicles should be commended for their choices – as well as having something different and unconventional for us all to admire, it also gives another dimension to the image of a 'classic car'.

Invariably, when first considering ownership of a classic car there will be a selection of marques and models that comfortably occupy the top ten choices. However, as shown with the owners' cars on the following pages, the many good reasons telling you to acquire something conservative are often outweighed by a single compulsion to own something quirky. The reasoning behind these apparently brave choices will be explained, ranging from giving a rare car a deserving second chance to enhancing a lifestyle, or just purely offering another outlet for a restorer's hobby.

But for all those who own one of these 'classic eccentrics', perhaps there is one motive they all share – they can be assured that, more often than not, no one else has anything quite like their car.

goggomobil ts300 coupe

'When a relocation to England beckoned for Dirk Wurm, selling his 1967 Goggomobil TS300 Coupé was not an option. The only workable solution was to air-freight the tiny Bavarian wonder in a Boeing 707 cargo plane from Frankfurt to England – despite the freight costs far outweighing the car's value. The car had been good to him; it was his first car and had served him well throughout his student days, so to just disown it was selfish,' explains new owner Paul.

'I first saw the car for sale 15 years ago but missed out on buying it. Then five years ago I spotted it for sale again – I was not going to miss out a second time.

'When the 300cc, two-cylinder, two-stroke, air-cooled engine starts up with full choke it smokes like a barbecue. When a hill is unavoidable the 15hp engine makes a noise like 50,000 angry wasps with plumes of blue smoke vortexing out the back like a vapour trail on a fighter jet! Quite honestly the driving experience is like nothing I have experienced before. Top speed 53mph – trust me, that's fast enough. It's easy to love the Goggomobil. It's like a faithful old dog – it struggles at times but eventually it will get there. Just a gentle tap of appreciation on the bonnet is enough of a reward.

'I attend classic car shows with my driving companion, Manfred von Geisdorfer, a life-size human skeleton dressed as an airman. It's fair to say the car is eccentric and also fair to say that I embrace the eccentricity.'

7600 PZ

car notes

After WWII Andreas Glas, originally a maker and repairer of farm machinery, sidestepped into producing motor scooters and then cars under the marque Hans Glas GmbH. From their factory in the Bavarian town of Dingolfing they produced the quirky Goggomobil TS Coupé microcar.

The air-cooled, rear-engine vehicle was introduced at the 1957 IFMA show (International Bicycle and Motorcycle Show). The definition of a microcar varied from country to country but the aim was to take advantage of the tax and licensing regulations for low-cc engines. For such a small car, it is packed with many idiosyncratic details both inside and out,

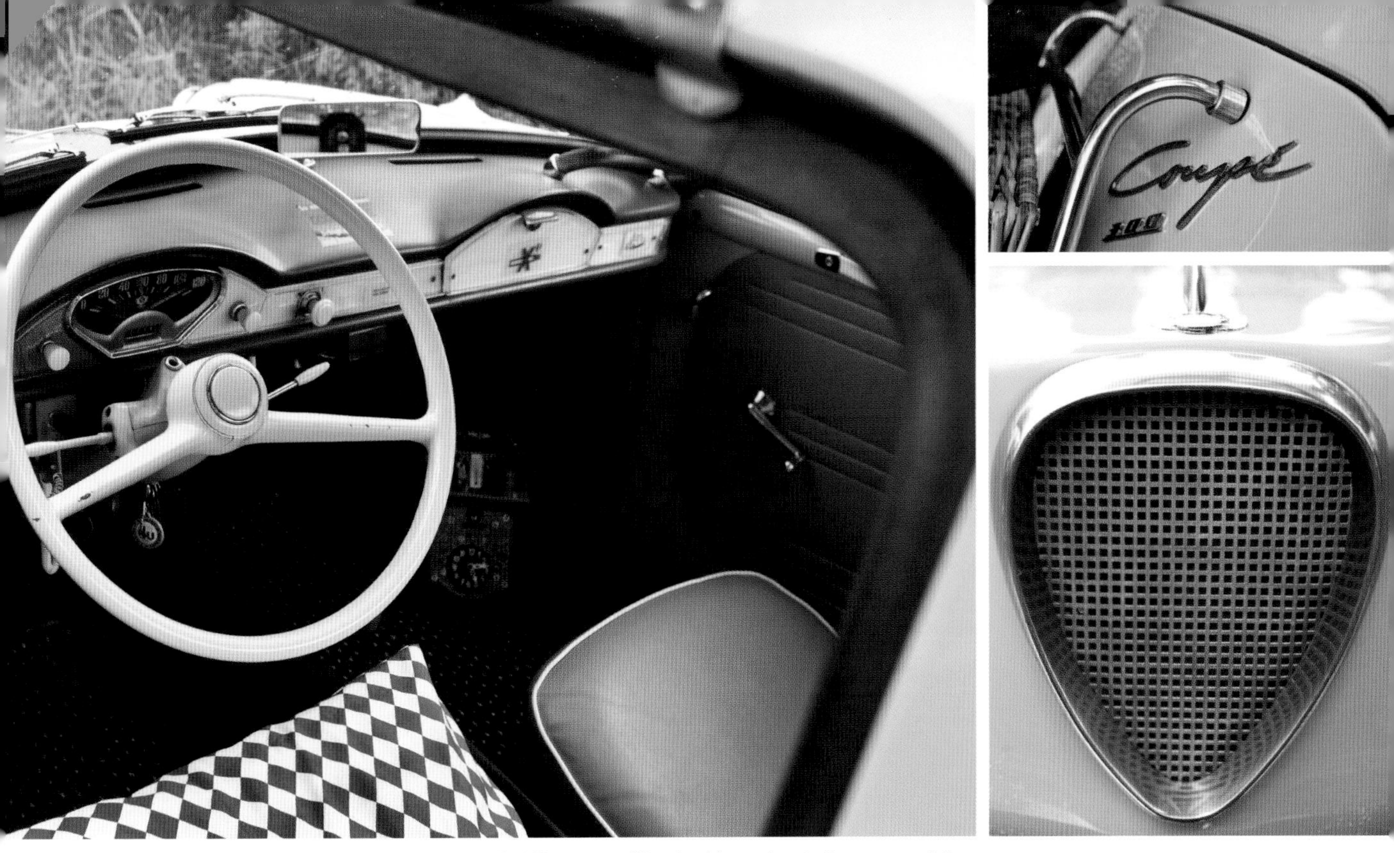

some standard and some retro-fitted: fishlike rear gills, dashboard switches resembling organ stops and an array of exterior styling more likely to be seen on a car twice the size and price.

Despite the TS Coupé costing 10 to 20 per cent more than its sedan counterpart, over 66,500 were produced in total. In 1964 Hans Glas introduced the 1300 GT Coupé designed by Pietro Frua. After the company acquisition by BMW in 1966 (as a way of accessing Hans Glas' valuable patents) the Goggomobil was continued and renamed the BMW 1600 GT. As of 2011, BMW's largest factory is still at the same Dingolfing location, although in new production facilities.

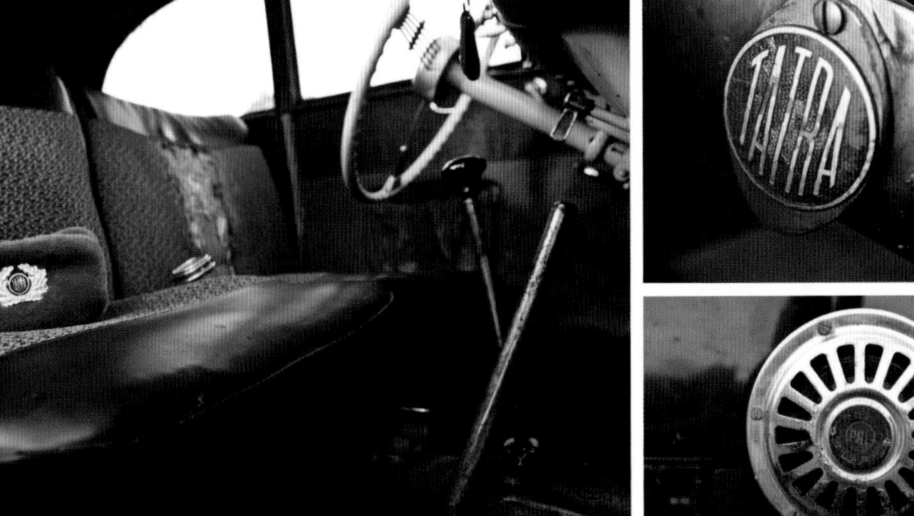

tatra t97

Resembling a cross between an automobile and 'The Rocketeer', Ian's 1938 Tatra T97 is a head-turner and conversation-starter for those keen to know more.

'My preoccupation with Tatras started as a result of my father cutting out unusual and interesting pictures of cars from magazines and newspapers, which he would then put into a scrapbook for me to look at. One of the cars I could readily identify at three years of age was the Czechoslovakian Tatraplan T600.'

These early recollections of cars stuck with Ian, especially the aerodynamic Tatras of the 1930s. But it was his wife, Kirsten, who eventually seized a chance opportunity to set them on the path to ownership.

'In the early 90s, while working in London, Kirsten was introduced to a Czech businessman called Jan Kašpar. Kirsten mentioned that I was fascinated by Tatras. Jan looked blankly at her and said, "Why would he be interested in Tatras?" She explained and since then Jan has become a good friend of ours, having helped us to buy our first Tatra. However, we still wanted to fulfil our original plan of owning a finned Tatra.

'Having founded Tatra Register UK in 1993, I'd long been aware of two T87s in Wales, both in need of restoration, and when they came up for sale over a decade later we bought them, despite the work needed. Shortly afterwards, though, an opportunity arose to buy an up-and-running T97. Since then the car, or as my wife refers to it, "the little sweetie", has served us well. We drive it everywhere and have even attended a German Tatra gathering in it. Unlike most of the other cars that were present we arrived without the use of a trailer, and the T97 behaved itself impeccably.'

car notes

The Tatra is often controversially claimed to be the inspiration for Ferdinand Porsche's VW Beetle. Designed by Hans Ledwinka and Erich Übelacker, the T97 was actually an expensive, largely hand-built car based on the company's V8-engined T87, advanced in engineering and design, especially considering the Tatra 97 was built in 1936, only 33 years after the first Ford Model A. The dramatic fin probably had little if any practical benefit, but featured on many contemporary experimental aerodynamic streamliners in the early 1930s, and on all Tatras from 1934 to 1953. Most manufacturers chose to take the conventional route, leaving Tatra to carry the flag for rear-engined, air-cooled streamliners. It boasted a top speed at the time of 82mph, which most British family cars wouldn't match until the 1960s. As well as suicide doors (i.e. rear-hinged), the unconventional styling continued on the inside with an anti-clockwise speedo. Just 508 T97s were built before WWII stopped play. The company continues to build rugged off-road trucks but stopped making cars in 1998 when it was, at the time, ranked the world's third oldest car manufacturer after Mercedes-Benz and Peugeot.

willys jeep

'As a young lad of no more than six, I remember Hollywood rolling into my home town of Steeton – which was being used as a location for the 1970s WWII film *Yanks*. My mum took me to a local garden party where on show were some of the vehicles used in the film. It was fantastic, it was like the GIs were back in England with jeeps, trucks and soldiers everywhere. Lucky ones like me were taken for a trip around the grounds in a jeep!' explains Jonathan. His interest in military vehicles, and especially jeeps, was rewarded when one Christmas he was given a toy Tonka jeep by his parents.

'I love music generally, I play rhythm guitar in a covers band, but the 40s era rubbed off on me in other ways as my grandad introduced me to big-band music. My girlfriend and I were at a 40s-themed dance and I noticed a Willys jeep. I mentioned to her that I'd always wanted one...her reply was, why not get one? So I did! Not only is it fun to drive but the dog loves a trip in it too.

'I had totally forgotten about my childhood toy jeep until recently when my parents found it and returned it to me – it's funny how things do a full circle.'

3
2
49
494
KP

car notes

Since WWI attempts had been made to mechanise army forces. In 1940 the US Army asked 135 US automotive manufacturers to submit a design based on exacting specifications: a general-purpose, personnel or cargo carrier especially adaptable for reconnaissance or command, and designated as a ¼-ton 4x4 truck.

Speed was of the essence: 11 days were given for bids to be received, then 49 days to deliver a prototype and a further 75 days to produce 70 test vehicles. Of the three companies that bid, Willys-Overland Motors and Ford ultimately

received the contract. Years of legal wrangling finally saw Willys win the right to use the term 'jeep' in 1950. Over subsequent years Willys introduced four-wheel drive to the public with its CJ (Civilian Jeep), making this the first mass-produced 4x4 civilian vehicle. The trademark slated radiator grill was designed by Ford and made standard with both manufacturers in 1942. Even today Jeep (now a marque of Chrysler) retains the distinctive grill featuring a standard number of vertical openings. During WWII 363,000 jeeps were produced under the Willys name and 280,000 under Ford.

porsche 356a

When, in the late 1950s, the Austrian police force found themselves being outpaced by criminals in faster cars, the powers that be took the opportunity to obtain four Porsche 356A convertibles – 'cops and robbers' in style! 'I first spotted the 356A when my current Porsche was at the garage. There I spotted a familiar shape contoured by a dust sheet,' explains Helen, a self-confessed Porsche-aholic with a very understanding husband. 'As I removed the sheet I saw my hunch was right; there stood a 356, in a good but unloved condition. I was further intrigued why an Austrian police uniform was in the rear of the car. I was put in touch with the owner of the car who happily explained the full history, and then after some tough negotiations I managed to buy it. With already having four Porsches and owning 18 before that, I felt that purchasing another one would not be such a big hurdle to overcome.

'I wanted to restore the car back to exactly the same specification as when it was in service. The 356A, now complete, has been on many journeys, including a memorable trip back to the streets of Austria. I can't help but imagine the rivalry between officers all those years ago when deciding who was going to drive the cars that day.'

car notes

In 1948 the first Porsche 356 'No. 1' prototype was created by Ferdinand 'Ferry' Porsche (son of Dr Ferdinand Porsche, the company founder). The original 356 body styling was the work of Porsche employee Erwin Komenda.

Like the Volkswagen Beetle (designed by Porsche senior), the 356 was a four-cylinder, air-cooled, rear-engine, rear-wheel-drive car, and there were accents of the Beetle in its design too. The 356 was the first production car produced by Porsche, marketed as an agile, two-door sports car available in coupé and, later, convertible models. The 356 design was the signature starting point that future Porsche designs would follow, including its successor, the 911.

Initially it used many Volkswagen parts for economy but as time went on the parts sharing was reduced. In the first two years of production Porsche manufactured just 50 cars but, with respect growing for the 356's handling and high quality, sales increased, with owners looking to race their 356 as well as use it for a daily drive. In 1964 over 10,000 were ordered, and by the end of production in 1965 a total of 76,000 had been sold.

MAC
89327
MAC
DMC
SIJ 1079

delorean

'I bought my factory-original, pre-production, right-hand-drive DeLorean in 1997. The car started its life at the Dunmurry factory as a test car until the collapse of the DeLorean business in 1982,' comments Michael. 'It was then sold by the receiver to a gentleman who wanted the iconic car to promote his stainless-steel company.'

After serving its purpose, the DeLorean was left in a garage to deteriorate. Just in the nick of time it was rescued, restored and brought back to its present condition. 'It's a fantastic car to own and often attracts a crowd of admirers. I take great pleasure in allowing people to have their photograph taken with the car – few other cars evoke such a reaction as the DeLorean! Another plus is the social aspect; it gives owners the opportunity to collate and swap car-related tips, helping them keep their cars in tip-top condition – after all, there are no dealerships available to call upon. Ownership has taken us on trips to America, France and Holland, and to Eurofest in Belfast, with future trips planned.'

Michael's DeLorean retains its shine through somewhat unconventional cleaning techniques. He explains, 'Through experimentation I have found that cleaning the car with washing-up liquid followed by a wipe over with furniture polish gives the best results. Not your normal method, but then again this is not your average car.'

'Oh, and I've never driven the car at 88mph...just in case you were wondering.'

car notes

John DeLorean's career in automotive engineering led him to General Motors before he left in 1973 to start his own business, the DeLorean Motor Company. The DeLorean DMC-12, which took five years to go from prototype (1976) to production (1981) – was aimed solely at the American consumer. It was the only car the company ever made before it went bankrupt in late 1982, by which time approximately 9,000 DMC-12s had been produced. *The Back to the Future* film that immortalised it in 1985 came too late to save it.

The sleek, two-seater sports car with its trademark gull-wing doors (evident by the unique dashboard 'door open' warning light) and brushed stainless-steel exterior was the

design of Giorgetto Giugiaro, an Italian designer responsible for a host of iconic supercar designs as well as more mainstream vehicles including the VW Golf and the Fiat Panda. Renault supplied the engine, with the chassis and bodywork details engineered by Lotus.

The cars, originally intended to be made in Puerto Rico, were built in Dunmurry, Northern Ireland, after the Northern Ireland Development Agency offered financial incentives to the company. In 1995 an entrepreneur called Stephen Wynne acquired the DeLorean Motor Company name and subsequently the trademark of the DMC logo. Today, made-to-order DeLoreans can be assembled from new original stock parts.

morris 1800

'For about 20 years I had noticed a redundant rally car gradually deteriorating in someone's driveway. If it hadn't been for my existing restoration projects I would have rescued it sooner,' explains Ian. 'Without a clue about its history I approached the owner and asked whether he would consider selling it, which he was very eager to do. He briefly explained the history of the car, which further confirmed that its current condition was not befitting a car with such pedigree – making it even more worthwhile restoring.'

Driving for 39 days, about 16,000 miles and over some of the harshest terrain in the world was not a problem for three plucky ladies in 1970, who took it all in their stride and competed in the London to Mexico World Cup Rally. Jean Denton, Liz Crellin and Pat Wright, all experienced ralliers, were selected by *Woman* magazine to take on the challenge of a lifetime.

Pat continues, 'This Morris 1800, also known as the Landcrab, was prepared for us by the Special Tuning Department of British Leyland. We started off in London as Car 91 and had a fairly uneventful journey through Europe – only managing two hours' sleep in Sofia, Bulgaria, the first rest since leaving London. At Lisbon, after 4,500 miles, we were in 26th place. We resumed in Rio de Janeiro and by the halfway point, Santiago, had moved up to 24th – by now 43 cars were still in play. On to Lima and despite episodes of altitude sickness (remedied with coca leaves) we made it over the stunning Andean high passes and then slogged up the Pan-American Highway. We finished at Mexico City in 18th place out of only 23 finishers. We clocked in, parked the car and plunged fully dressed into the hotel swimming pool! This challenge certainly earned itself the title "The World's Toughest Rally".'

motorwoman
JEAN DENTON
PAT WRIGHT
LIZ CRELLIN
COFFEE
CAFÉ
91
DAILY MIRROR
WORLD CUP RALLY
the beauty box

BATTERY CONDITION
VOLTS
s/h lights
wiper
spots
halda
TWINMASTER
HALDEX A.B.
HALMSTAD
SWEDEN
COLD
BOOST
RAM
CAR
RAM
BOOST

BRITISH LEYLAND
Willie
WORLD CUP 1970 SUPPORTER

car notes

The road-going version of the Morris 1800 (also known as the Austin 1800 and Wolseley 18/85 or 'Landcrab') was produced by the British Motor Corporation (BMC) from 1964 to 1975. Credibility was earned with rallying success, second in the 1968 London to Sydney and claiming three of the top twenty places in the 1970 London to Mexico.

The engineering was ahead of its time and the suspension delivered an enviably smooth ride. The exterior was a collaboration between Alec Issigonis and Italian styling house Pininfarina; its huge expanse of glass was a departure from the norm in 1964. The spacious interior was adorned with leatherette, imitation wood and some chrome detailing, complemented by a functional but unusual dashboard (featuring a ribbon speedo) and very nearly a front bench seat.

Despite everything pointing to a success like its predecessors the Mini and Austin 1100, sales struggled (reaching 386,000 across all marques) and in 1975 it was replaced with the wedge-shaped Princess.

ford model t

'As the story goes, soon after manufacture my 1926 Ford Model T was exported from Canada to New Zealand,' explains Peter. 'What happened to it after that is uncertain, so quite how or why it ended up, very dishevelled (or shabby chic as I refer to it), suspended on a mezzanine platform in a workshop is difficult to conclude. When a friend of mine first came across the car in New Zealand and enquired about buying it, he was told by the mechanic in no uncertain terms, "If I bring it down you take it – I'm not putting it back!" So with the help of a fork-lift truck it reluctantly came down, the chassis becoming slightly bent in the process. My friend then shipped it halfway around the world to England in 1987 and I purchased it from him in 2002.

'Despite the stuffing oozing out of the upholstery, a hood that is better left alone, the odd patch of rust and a minor case of woodworm she is in good condition. She might look like an old jalopy, but mechanically the car is fine and during the summer months it gets me to and from the pub I own with relative ease. It has no speedo, not that it could ever reach or break any speed limits, but to be on the safe side I rely on the speed readout on my satellite navigation – how about that for old meets new! You may laugh, but it's not as easy as you might think to keep a car like mine in the condition it is. Most repairs go unnoticed, as do new parts, as they first need to be manually distressed and aged to avoid looking out of place!

'She got the nickname "Old Girl" because when I arrived at the pub one day with my wife a regular customer said, "I see you've got the old girl with you." I replied, "Yes, and I've also brought the car!"'

Ford

car notes

1908 was a landmark in motoring history when Henry Ford started production of the Model T or 'Tin Lizzie'. Henry Ford's motivation was for a quality, affordable car so that many, not the privileged few, could experience the joy that motoring brings.

The first month of production was slow – just 11 cars. Factory improvements were made and by 1910 nearly 12,000 Model Ts had been built, but still not enough to satisfy demand. The car was intentionally very versatile, making it able to fit in with everyday realities; even with wooden-spoked wheels it was drivable over all manner of terrain. The removal of a wheel and addition of a pulley could turn the T into a conveyor belt, thresher or bucksaw. Until the introduction of electric lighting, early Model Ts were

fitted with acetylene gas headlights and kerosene side and tail-lamps. After moving to a new factory Model Ts were being built in 93 minutes (as opposed to a leisurely 12½ hours previously) and one rolled off the production line every three minutes.

When Ford had built its 10-millionth car it was said that of all the cars in the world, half were Ford. In 1925 the streamlined factories worldwide were producing 10,000 cars a day. By now other manufacturers had started to produce superior cars and Model T sales suffered. On 26 May 1927 Henry Ford watched the final Model T of over 15 million roll off the assembly line – this was truly the first globally successful mass-produced automobile.

citroën ds pallas

When buying a classic car your worst nightmare would be the car not starting once you had handed over your hard-earned money. This was the reality for Pete when he went to collect this 1975 Citroën DS Pallas. Pete explains, 'Instead of the seller making a hasty retreat his reaction, with a cheeky smirk on his face, was "Well, this will give you the chance to learn how to repair it, won't it?" With that the bonnet was up and moments later I was up to my elbows in grease, being given verbal instructions whilst being passed tools.'

Although an unconventional start to owning a car, Pete admits it has stood him in good stead and the DS, called 'Yvette', is mechanically fine. So much so that he and his wife Nichola have twice embarked, without even a murmur from the car, on a 2,000-mile round trip to their holiday home in Italy. 'Italians have a love affair with the Citroën DS, which manifested itself on our recent trip when we found a series of begging notes asking us to sell the car.

'I love design classics and can often be found rummaging at a car-boot sale early on a Sunday morning. When it comes to iconic cars Citroën have had their fair share – my personal favourite, of all the types I have had the privilege of owning, being the DS.

'The exterior is a little tatty round the edges – a life on the busy streets of London doesn't help – but as long as she gets us from A to B that is all that matters.

'The ride in a DS is an experience. The unique hydraulic suspension gives you a "floating on air" feel, which could leave some feeling a bit seasick, but when we are on one of our long journeys you won't find us complaining.'

car notes

Some things are not worth rushing – witness the Citroën DS (pronounced 'Déesse' – goddess), which remained in secret development for 18 years. The DS was designed to be the successor to the Citroën Traction Avant. Its launch at the 1955 Paris Motor Show was something of a record, as it received 743 orders in 15 minutes and 12,000 by the end of day one.

The futuristic body (styled by Flaminio Bertoni and aeronautical engineer André Lefèbvre) has long elliptical lines with conflicting angles that shouldn't work, but do so beautifully, as do the high-level rear indicators mounted in champagne

flute-like enclosures. The DS was a beacon of French recovery in the post-war years. Technological advances included a glass-fibre roof, power steering, semi-automatic transmission, hydropneumatic suspension, an automatic levelling system with variable ground clearance and, in 1967, directional headlights that swivelled up to 80° as the driver steered.

The 20-year production run that followed the 18-year development proved that if something is worth designing, then design it right. This was underlined when the DS was placed third in 1999's prestigious Car of the Century award.

sports

How do you define a sports car? The mind's eye will inevitably conjure up simplistic, almost childlike shapes of cars with sleek lines and long curves. Ask five different people for their definition of such a car, however, and invariably you will get five conflicting views. The truth is that there is no clear-cut definition which encompasses the term sports car. Inevitably the clichés of top speed, horsepower, handling and engine size will come into the equation somewhere, and these are certainly valid, but surely the most important factor is the experience it rewards the driver with.

Some may balk at this unembellished view, but no classic sports car owner would deny that the exhilaration of driving such a car is difficult to contain. These outbursts of exuberance will most likely be part of the reason that many others will be spurred on to acquire their own dream sports car.

Among the fascinating examples in this section are a cherished family-owned Mercedes, well used rather than closeted away, the man who now lavishes his attention on a Porsche rather than a horse, and the Triumph Stag owner who could no longer endure the 'what if' prospect of never fulfilling his dream.

When driving these cars it is often a case of having no particular destination or objective in mind. So put aside any worries when considering the practicalities of owning a sports car – ownership is purely about the driving and the enjoyment it brings.

mercedes 300sl roadster

'My late parents were an amazingly eccentric couple who adored each other but were polar opposites in personality and always playing practical jokes on each other,' laughs Martin. 'My father, Ron, a very placid man, was English and my mother, Trudi, who had a fiery temper, was from Stuttgart, Germany (home of Mercedes).' Martin's father, a self-taught engineer, purchased the 300SL Roadster in 1963. His passion for Mercedes began then and never ended. Martin continues, 'I guess for me the passion started when, as a child I was taken on some amazing European road trips. The car is only a two-seater, with a small space behind the front seats that I would just about fit in. Occasionally my father would take the car out for a so-called "short spin", only to return several weeks later having driven to Greece and back. As a result of all these wonderful adventures the car is now closing in on 400,000 miles.

'My father was never happier than when tinkering on the car in his custom-designed three-storey garage. If a car part did not exist, was too expensive or delivery was too slow he'd just make an exact copy! His knowledge and love of Mercedes cars led him to be on first-name terms with the top brass at Mercedes HQ. Even when my father became very ill he would sit with me in the workshop, watching and doing his best to convey his opinions while I worked on a car.

'I'm left with a tough legacy to uphold – the cars I cannot part with and a house like a museum full of my parents' idiosyncrasies. However, my plan one day is to pack up and move with the cars to the Black Forest, a special place for me and also a wonderful location to experience the thrill of driving the roadster.'

300 SL CLUB e.V.
MERCEDES-BENZ 300SL CLUB e.V.
36
26-30 September 2001
34
1981
2004
60
MERCEDES-BENZ 300 SL CLUB e.V.
KITZBÜHELER
1997
Rally
6
1979
19
MERCEDES-BENZ 300SL CLUB e.V.
1996
Classic
4 PK

WT 48

car notes

Max Hoffman, a prolific Austrian-born American car distributor, was well known for suggesting cars to the manufacturers which he thought they should develop to fulfil the needs of cash-rich America. His request to Mercedes-Benz was for a road-legal version of the Mercedes W194. It was duly acknowledged and it fell to engineer Rudolf Uhlenhaut, from the Mercedes-Benz racing car department, to make the adaptation.

The stunning 300SL (*Sport Leicht*) was launched at the 1954 New York Auto Show. The Roadster, introduced in 1957, replaced the Gullwing and remained in production for a further six years. With conventional doors it avoided the undignified way you had to enter or exit the Gullwing when negotiating the very high sill. The sleek profile is enhanced by way of slanted heat vents on the front wings and 'eyebrow' mouldings over the well arches. The contrast between the car's power and elegant interior styling is striking – the pencil-thin gear stick appears inadequate for such a potent engine – but this is part of its charm, as there is no trade-off in style over performance.

As with any supercar, build numbers were low, creating iconic status; only 1,400 coupés (Gullwing) were made and 1,858 Roadsters.

mga mk 2 coupe

'I'm a self-confessed MG nut,' admits Paul. 'My first was an MGB roadster, which I bought in 1989, I then added a 1947 MG TC (supercharged) in 1999 – such an exhilarating drive! It was whilst I was looking to buy a T-type that my attention was sidetracked by a stunning, seductive, sleek MGA coupé in a showroom – you can't fail to appreciate the lines of the MGA, which is considered by many enthusiasts to be the prettiest car the company produced.

'The TC and MGB were fantastic, but as they are both soft-tops use was hampered in poor weather. I considered other MGs and, dare I say, manufacturers, but kept coming back to the MGA coupé. My heart was set on the desirable Mk 2, ideally with a body colour of Iris Blue, typifying the MG era of the 50s and 60s. I sold both the TC and MGB but, not wanting to be car-less for long, seriously started looking for the MGA. Quite fortuitously, in 2008, this MGA coupé came up for sale in Wales. It drives as good as it looks and is used extensively all year round – including trips to the Le Mans Classic and a 2,000-mile rally around the Western Isles, as well as attending the Angoulême Ramparts race weekend.'

car notes

In 1955 the time was right for MG (a division of the British Motor Corporation) to move away from the 'comfort blanket' that was its 1930s-styled sports cars. Design clues for the MGA had been hinted at as early as 1951 when a re-bodied TD Midget took part in the Le Mans 24 hour race. When the MGA was officially launched at the Frankfurt Motor Show no one could have predicted such a bold new up-to-date design. It was such a departure that advertising declared it the 'first of a new line'. MG had well and truly bid farewell to the 1930s.

The MGA was available as a coupé and open-top roadster, both of which were true two-seaters and had beautiful flowing wings that curved their way from front to back. The MGA's success was assured: it was good to look at, robust, mechanically simple and reliable, and it rewarded the driver with good performance and handling – the perfect recipe for a true sports car.

During its seven years of production (ending in 1962) only about 6 per cent of the 101,081 built were sold in the UK, a record export ratio for a British car. The MGA is now considered one of the most beautiful sports cars of all time.

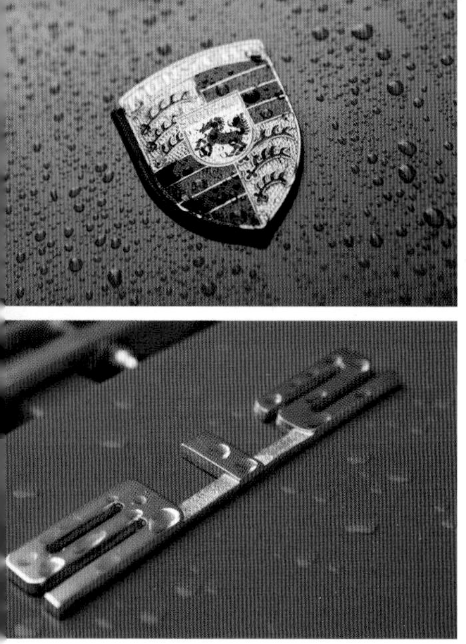

porsche 912

When for many years you have had an almost daily military regime to follow, the shock of any change can leave you somewhat lost. This was the reality for Paul, who took early retirement after 25 years' service as a mounted officer in the London Metropolitan Police.

'Each day my horse needed grooming, bridles polished, uniform pressed, all ready for inspection. So when the routine ended I needed something to fill the void, something else to obsess over if you like. Myself and my wife Sue have always appreciated classic cars, in fact Sue owns a superb 1947 Morris 8. However, for a long time my weakness has been the classic Porsche 911, but with prices heading sky high I plumped for the 912, which has all the looks and fun of a 911 without the paranoia of the rear end swinging out sideways!

'After some careful research I headed off with a deposit in my back pocket with the full intention of using it. The car that so confidently relieved me of my money was this immaculate 1968 Porsche 912 in Blood Orange, so typical of the rainbow spectrum of colours available at the time.

'The 912 is the first true classic car I have owned and it never disappoints. From the fantastic unmuffled throaty noise (accelerating out of third gear has to be my favourite tone) emanating from the air-cooled four-cylinder engine, to the perky handling which really comes into its own on a twisty back road – it really is a joy.'

PORSCHE
SRR 365F

car notes

When production of the Porsche 356 stopped in 1965 company executives were concerned that its higher-priced 911 replacement (which proved to be one of the most successful sports cars of all time) would be an issue. In 1965 a lower-cost entry-level model based on the 911's chassis and body shell was introduced – namely the 912.

It used the reliable four-cylinder engine from the 356, which still packed a sporty punch, along with the handling and appealing iconic swept-back teardrop silhouette and flared rear wheel arches of the 911. The lower cost was enticing to customers and initially 912 sales outstripped the 911. In 1970 the 911 platform was deemed financially stable and Porsche's eagerness to introduce a new model (the 914) made the 912 redundant.

Over the five years of production Porsche built 30,000 912s; it was reintroduced as a stopgap as the 912E for a short period in 1976.

volvo p1800e

'When handed the keys to my new company car, it's fair to say a Volvo wasn't my ideal choice. More desirable marques had been mooted, however my fleet manager was unpersuaded so a Volvo 740 estate it was.' Jonathan perceived Volvos as a bit stuffy and boring, but soon retracted that view after a few enjoyable days behind the wheel. In fact, he was so impressed that when unfortunately made redundant he bought the car for his personal use.

Jonathan continues, 'I've always had a passion for cars, I spend a considerable time doing what my wife calls "playing" with cars. So like many, I decided that on retirement I would indulge myself and buy a classic car. I first considered MGs and Austins – cars from my childhood that my father and grandfather drove. But I was intrigued to find out what classics Volvo had to offer, especially the often overlooked Volvo sports car – the P1800. It is one of those unexpected classics that surpasses expectations. What attracted me most is that its design is such a break from most people's perception of a Volvo. So with a buyer's guide in hand I went in search of one and luckily soon found this fantastic 1971 example.

'When my son learned of my acquisition he was sceptical and, having no real interest in cars, he was perplexed by the words "Volvo" and "sports" being used in the same sentence. However, on closer inspection, like myself, he too has changed his mind and I have to say he now uses it more than I do. It may not be the fastest sports car, but what it lacks in speed it makes up for in looks.'

GPP 8K

1800
E

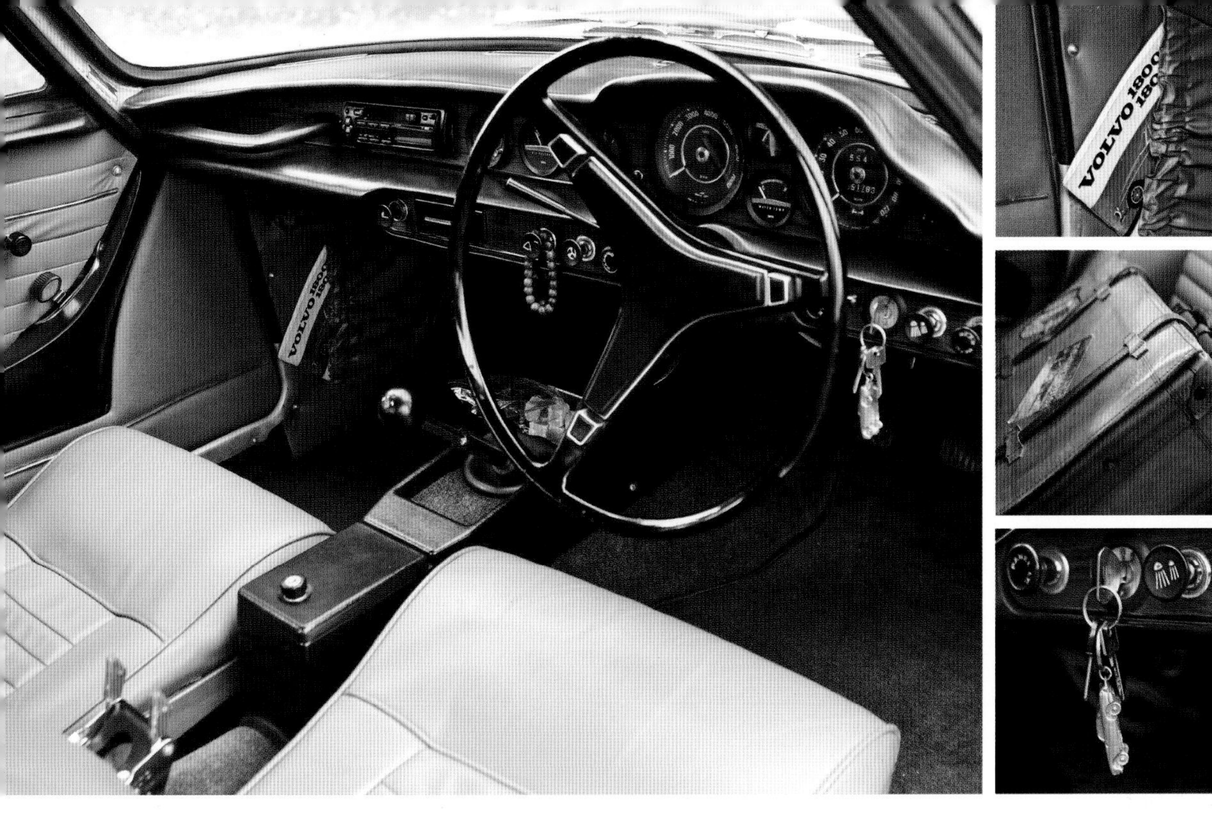

car notes

Despite their first attempts at a sports car being a disaster (namely the P1900 selling only 68) Volvo carried on undiscouraged and in 1957 started the P1800 project. They employed the talents of Pelle Petterson, an award-winning designer and student of Pietro Frua. Pelle, from Stockholm, was also renowned for his sailing boat designs. These maritime skills, one could argue, were reflected in the classic, elegant, clean lines of the P1800, in particular the rear fins, resembling those of a hydrofoil boat, and the door handles flowing seamlessly into the rear wing. Scandinavian influences are visible with an interior that straddles the divide between trend and timelessness, accompanied by an overall functional yet simple design.

Production began in 1960 at Jensen Motors, England, until the contract was terminated (after 6,000 cars) due to quality issues. It then relocated to Gothenburg and was renamed the P1800S ('S' denoting assembled in Sweden). After several incarnations the P1800's performance matched its sporty look, with the fuel-injected 1800E. At its inception this was a bold-looking creation, and raised a large amount of attention. However by 1971 the design was looking dated; production ceased in 1973 with 39,407 P1800 coupés having been built.

triumph stag

'Life is too short, and after a year of very unfortunate news concerning close friends I made a rash, but pragmatic, decision to buy something I had loved and wanted for years. I didn't want a situation to arise of me saying "what if" when too old to do anything about it,' explains Dave. 'It was a Saturday morning, I had already made some rumblings to my wife about a Triumph Stag and she had given the green light to proceed, so I wasted no time. Having already circled a few possibilities in the classifieds I felt one in particular, a 1974 V8 Stag, warranted a viewing.

'I told my wife I was just popping out to "look" at the car in question and to that she jokingly remarked to her mother that I would probably come back with the car...she wasn't wrong.

'Having worked for many years in the motor industry, at a well-known factory in Dagenham, I have great respect for the work that goes into building a car, but at the end of the day it is a lump of metal which is built to be used, and even though I love it to bits it's not something to be precious about. One treat I do relish, though, is driving through a motorway tunnel with the roof and pedal down and the V8 warbling – who needs a radio?'

car notes

In 1964, at designer Giovanni Michelotti's request, Harry Webster, Triumph's Director of Engineering, supplied him with a Triumph 2000 so he could experiment and develop a car to promote his skills at the Turin Motor Show. It was, however, on the understanding that if Harry liked the resulting car he would have first refusal to put it into production.

Giovanni envisioned a luxury two-door, four-seater convertible sports car. To Harry, what Giovanni had created was a car that could help launch Triumph in America. Two unique aspects of the car's design were the elliptical shapes carved into the front and rear wings, making them almost symmetrical, and the distinctive T-bar roof.

Initial response was good after its 1970 launch, but problems with the V8 engine plagued almost every model. For British Leyland (owner of Triumph) the car was neither a great success in England nor America, with around 25,000 made between 1970 and 1977. In recent years the Stag has gained the respect it originally sought, being regarded as a comfortable, well-refined and well-appointed sports car.

morgan 3 wheeler

Reverting to three wheels when everyone else seems to be getting along just fine with four could be deemed as a retrograde step; unless, of course, you've experienced first-hand the fun of the Morgan 3 Wheeler. The driving experience Morgan are striving for is a true seat-of-your-pants, back-to-basics, smile-a-mile machine! (Sadly, a scantly made claim these days.) Morgan, having last manufactured a three-wheeled vehicle in 1953, have revived it – truly a future classic. When harnessed into the open-top, fighter-plane style cockpit you immediately feel at one with the vehicle and could oh so easily imitate a rat-a-tat-tat machine-gun sound effect. Such immaturity – which isn't helped by the engine starter button being that of the Eurofighter's bomb hatch release! The engine doesn't just start...it roars into life – instantly evoking feelings of being a vintage racing driver on the starting grid.

I defy anyone to walk away from a drive in the Morgan 3 Wheeler without loving it and feeling years younger!

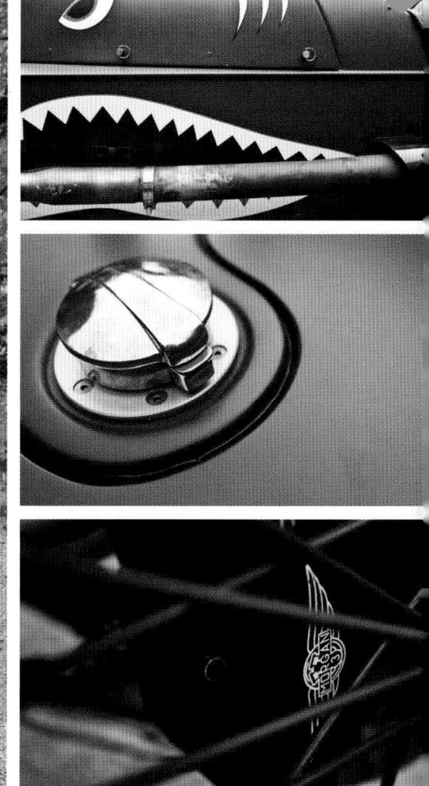

car notes

In 2011 Morgan announced the resurrection of an automotive design from a bygone era and created an aviation-style experience solidly on terra firma.

The exposed engine, chassis and exhaust pipes could all too easily be covered, but doing so would change the characteristics of the vehicle. The key to the 3 Wheeler's success is the driver's engagement with the elements. The fuselage provides just enough protection from the elements but not too much to disengage the driver. The vinyl graphics give a sense of fun and personalisation and a modern edge to an otherwise traditionally inspired vehicle. The padded leather cockpit further inspires aviation tendencies but adds refinement at the same time.

If you're tempted, you might have to wait a while – the new 3 Wheeler's complete first year of production has already been snapped up and pre-ordered by eager enthusiasts.

sourcebook

owners clubs

AMC Pacer
www.amcpacer.com

Austin Allegro
www.allegroclubint.org.uk

Austin 7
www.austinsevenownersclub.com

Bentley
www.bdcl.org

Citroën 2CV
www.2cvgb.co.uk

Citroën DS
www.citroencarclub.org.uk

Corvette
www.corvetteclub.org.uk

DeLorean
www.deloreans.co.uk

Facel Vega
www.facelvegacarclub.co.uk

Fiat 500
www.fiat500club.org.uk

Fiat X1/9
www.x1-9ownersclub.org.uk

Ford Capri
www.caprimk1ownersclub.com
www.capriclub.co.uk

Ford Model T
www.modeltregister.co.uk

Goggomobil
www.micromaniacsclub.co.uk
www.glasclub.de

Holden
www.holdenuk.co.uk

Jaguar E-type
www.e-typeclub.com
www.jec.org.uk

Mercedes-Benz
www.mercedes-benz-club.co.uk

MGA
www.mgownersclub.co.uk

Mini
www.britishminiclub.co.uk
www.miniownersclub.co.uk

Morgan 3 Wheeler
www.mtwc.co.uk

Morgan
www.mscc.uk.com

Morris 1800
www.landcrab.net

Morris Minor
www.mmoc.org.uk

Mustang
www.mocgb.net

Panhard
www.panhardclub.co.uk

Plymouth Chrysler
www.chryslerclub.org

Pontiac Trans Am
www.poc-uk.org

Porsche
www.912register.co.uk
www.porsche356registry.org
www.porscheclubgb.com

Renault
www.renaultownersclub.com

Rover P6
www.p6roc.com

Rover SD1
www.roversd1club.net

Tatra
www.tatra-register.co.uk

Toyota Celica
www.celica-club.co.uk

Triumph
www.tristagreg.org
www.club.triumph.org.uk

Volkswagen Beetle
www.vwocgb.com

Volkswagen Golf
www.vwgolfmk1.org.uk

Volvo
www.volvoclub.org.uk

Willys Jeep
www.willys-mb.co.uk

car care and accessories

Autoglym
Specialist car cleaning products
www.autoglym.com

Carcoon Car Cover
Protective airflow car storage system
www.carcoon.co.uk

Leacy Classics
Classic car parts for MG, Triumph, Mini, Morris Minor and Austin
www.leacyclassics.com

Retro Classic Car Parts Ltd
Classic car parts and quality motoring clothing
www.retroclassiccarparts.com

US Automotive
American car parts supplier
www.usautomotive.co.uk

car sales

Claremont Corvette
Corvette sales, service and parts
www.corvette.co.uk

Frank Dale & Stepsons
Rolls-Royce and Bentley sales and servicing
www.frankdale.com

Eclectic Cars
Eclectic car sales and classic car servicing
www.eclecticcars.co.uk

Maxted-Page & Prill Limited
Historic Porsche sales and service
www.maxted-pageandprill.com

Morgan Motor Company Limited
www.morgan-motor.co.uk

Morgan 3 Wheeler
www.morgan3wheeler.co.uk

car hire

Classic Car Club
www.classiccarclub.co.uk

Classic Car Hire
www.classiccarhire.co.uk

Vanilla Classics
www.vanillaclassics.com

museums

Beaulieu National Motor Museum
www.beaulieu.co.uk

Brooklands Museum
www.brooklandsmuseum.com

Haynes Motor Museum
www.haynesmotormuseum.com

credits

We would like to thank all the owners for allowing us to photograph their 'cool classic cars'.

All photography by Lyndon McNeil unless otherwise stated.
www.lyndonmcneil.com

beloved

Pages 12-15 VW Beetle, Lee Callaghan, London
Pages 16-19 Fiat 500, Christine Anderson, Buckinghamshire
Pages 20-23 Morris Minor, Budd Birkitt, Essex
Pages 24-27 Austin Seven 850, Peter Hayes, Buckinghamshire
Pages 28-31 Renault 4, Sandra Levet, Hertfordshire
Pages 32-35 Citroën 2CV, Chris Litchfield, Hertfordshire
Pages 36-37 Austin 7 Chummy, John White, Hertfordshire
Pages 38-43 Morgan 4/4, Tom Rickman, Cornwall
Pages 44-45 Triumph Herald, David Lewis, Surrey

retro

Pages 48-51 VW Golf, Roger Miller, London
Pages 52-53 Rover SD1 3500 Vitesse, Dennis Parker, Essex
Pages 54-55 Austin Allegro Vanden Plas, Chris Figg, East Sussex
Pages 56-57 HG Holden Ute, Ken Virtue, Australia (Photography by Hilary Walker)
Pages 58-61 AMC Pacer, Rebekah Hawthorn, Norfolk
Pages 62-63 Rover P6, Barrie Flemming, Middlesex
Pages 64-67 Toyota Celica GT2000, Duncan Hewitson, Essex
Pages 68-71 Ford Capri Mk1, Eddie Hughes, Hertfordshire
Pages 72-73 Fiat X1/9, Fiorenzo Coppola, Essex
Pages 74-77 Panhard 24 CT, John Passfield, Essex

glory days

Pages 80-81 Bentley R Type Continental, Ivor Gordon, London
Pages 82-83 Corvette V8 Speedster 'Duntov', Tom Falconer, Kent
Pages 84-87 Plymouth Belvedere, Paul Alford, Buckinghamshire
Pages 88-91 Jaguar E-Type, David Holland, Hertfordshire
Pages 92-95 Facel Vega, Bob Constanduros, West Sussex
Pages 96-97 Ford Mustang, Martin Cooper, Hertfordshire
Pages 98-101 Pontiac Firebird Trans Am, Steve Green, Buckinghamshire

classic eccentrics

Pages 104-107 Goggomobil, Paul Bussey, Hertfordshire
Pages 108-111 Tatra T97, Ian Tisdale, Oxfordshire
Pages 112-115 Willys Jeep, Jonathan Pittock, Essex
Pages 116-117 Porsche 356A, Helen Goff, Leicestershire
Pages 118-121 DeLorean, Michael Hooper, Hertfordshire
Pages 122-125 Morris 1800 Landcrab, Ian Feirn, Hertfordshire
Pages 126-129 Ford Model T, Peter Burdfield, West Sussex
Pages 130-133 Citroën DS, Pete and Nichola Jenkins, London

sports

Pages 136-141 Mercedes 300SL, Martin Cushway, Essex
Pages 142-143 MGA Mk 2 Coupé, Paul Camp, Hertfordshire
Pages 144-147 Porsche 912, Paul Davis, Essex
Pages 148-151 Volvo P1800E, Jonathan Woolf, Hertfordshire
Pages 152-153 Triumph Stag, Dave Barr, Essex
Pages 154-155 Morgan 3 Wheeler, Morgan Motor Company Ltd, Worcestershire

acknowledgements

I would like to thank everyone involved for their help in making this book a reality.

Thank you to Pavilion Books, in particular Fiona Holman and designer Georgina Hewitt for their continued support.

To Lyndon (this book's photographer) – your enthusiasm did not waver from the first to the last shutter release. Thank you Maureen Hunt and Richard Heseltine. To Emma Haddon and Sarah Bull for their help, support and for putting up with classic cars being the number one talking point for Lyndon and me for many months. Finally to my children Gracie and Imogen for taking up their summer holidays with classic car shows.

Most importantly, a huge thank you to the owners of the cars featured, who have not only given us access to their amazing cars but have also shared their stories, which has made this book an immense privilege to produce.

Both Lyndon and I would like to dedicate this book to our mothers, Maureen and Pauline.

chris haddon

Chris Haddon is a designer with over 20 years' experience. He has a huge passion for all things retro and vintage. Among his collection is his studio, which is a converted 1960s Airstream from where he runs his design agency.

Additional captions: page 1 corvette v8 speedster 'duntov'; pages 2–3 pontiac firebird trans am; page 4 tatra t97; page 6 plymouth belvedere; page 9 ford capri mk 1; page 10 austin seven 850; page 46 austin allegro vanden plas; page 78 plymouth belvedere; page 102 ford model t; page 134 volvo p1800e; page 160 renault 4

This edition is published in 2020 by
Pavilion
43 Great Ormond Street
London
WC1N 3HZ

A CIP catalogue record for this book is available from the British Library

ISBN 978-1-911641-56-8

10 9 8 7 6 5 4 3 2 1

Colour reproduction by Dot Gradations Ltd, UK
Printed and bound by Toppan Leefung Ltd, China

www.pavilionbooks.com